TOM BRADBURY

METEOROLOGY
AND
FLIGHT

A PILOT'S GUIDE TO WEATHER

THIRD EDITION

A & C Black • London

Published by A & C Black (Publishers) Ltd
37 Soho Square, London W1D 3QZ
www.acblack.com

First edition 1989; reprinted 1991
Second edition 1996
Third edition 2000; reprinted 2004

ISBN 0 7136 6831 8

Cover photographs © The White Planes Picture Company

A CIP catalogue record for this book
is available from the British Library.

Printed and bound in Great Britain by
Martins the Printers, Berwick upon Tweed

Contents

Conversion factors

1 m = 3.2808 ft   1 ft = 0.3048 m
1 mm = 0.039370 in   1 in = 25.40 mm
1 km = 0.62137 mile = 0.53993 nautical mile
1 mile = 1.6093 km = 0.86898 nautical mile
1 nautical mile = 1.8520 km = 1.1508 mile = 6076.1 ft
1 metre per sec = 2.2369 mph = 1.9438 nautical mph
1 ft per sec = 0.68182 mph = 0.59248 nautical mph
1 mph = 0.86898 knot = 1.6093 km per hour = 0.44704 m per sec
1 knot = 1.1508 mph = 1.852 km h = 0.51444 m per sec
1 km per h = 0.62137 mph = 0.53996 knot
1 kg = 2.2046 lb   1 lb = 0.45359 kg
1 oz = 28.350 grammes = 0.06250 lb
1 gramme = 0.035274 oz
1 tonne = 1000 kg = 2204.6 lb
1 UK ton = 1016 kg = 2240 lb

1 atmosphere = 1013.25 mbar = 101325 pascal = 760.0 mm Hg = 29.921 in Hg = 33.899 ft water =
    14.696 lb sq in = 10332 kg sq m
1 pascal = 1 newton sq m = 0.01 mbar
1 millibar = 100 pascal = 0.02953 in Hg = 0.000750 mm Hg
1 in mercury (Hg) = 33.864 mbar

1 kW = 1.3410 hp = 1.3596 metric hp (cv)
1 hp = 745.7 kW = 1.0139 metric hp

Introduction

This book is chiefly for those who fly for enjoyment and would rather look at the view than concentrate on the instrument panel. It is not intended as a text book on Meteorology, but as a guide to some interesting features of the weather which affect flying.

Everyone needs some background knowledge to interpret what is seen. Some pilots, especially those who fly both with and without an engine, seem to have an amazing ability to interpret features in the sky which many meteorological 'experts' do not see at all. What is observed depends to a large extent on how the observations fit into a mental framework. It is said that when sight was restored to a blind man he was shown an orange but was unable to recognise it until he had put out his hand to feel its roundness and texture. The eye could produce an image but the brain did not have the experience to interpret it.

This problem is not confined to the newly sighted. Some city dwellers do not notice what is happening in the sky above until a shower descends on them. Student pilots occasionally fail to recognise their home town when looking at it from above for the first time. Quite a few meteorologists only know a cloud by its statistical properties, not as a usable source of energy. Some landscape artists draw realistic clouds, while others put in mere splashes of colour to fill in the sky.

In any subject there are dull parts and complicated sections and weather is no exception. What one person finds tedious, another may find interesting. Sections which seem to be over technical can be skipped; the essential points are repeated where necessary in other chapters. This is why some items are to be found in several different chapters.

The appendices contain details of various Met. broadcasts which may be used to keep up with recent weather reports. These range from facsimile and radio teletype broadcasts, which require dedicated equipment, to plain language broadcasts.

Much space has been devoted to features of particular interest to those who fly without an engine. For these pilots even small movements of the air and subtle changes in the clouds are important. Pilots of powered aircraft with modern radio and electronic aids may, when they have to, go from take off to landing without seeing anything in between. To these experts weather is usually of secondary importance. Even so, there are some types of weather which everyone prefers to avoid, and one may need to fly to an airfield without VOR, ILS or radar.

It is hard to make any statement about the weather without adding qualifying phrases such as '. . . on most occasions . . .', '. . . provided that . . .', '. . . except when . . .'. This becomes wearisome to write and exasperating to read. However, the reader needs to be a little cautious in accepting any weather 'laws', especially when they are asserted with undue confidence. There are nearly always exceptions. Mathematical theories have achieved remarkable success in producing numerical forecasts but they cannot yet handle all the complexities of the weather. Honest observations are still important.

A note on units

As far as possible the units are those most often used in flying. These are an unscientific mixture agreed by the aeronautical authorities, who were constrained by the fact that many instruments were already calibrated in them when the subject was debated.

Speed is given in knots, height in feet, temperature in degrees Celsius. Distances are mostly given in nautical miles since this is convenient for navigation over a globe divided into latitude and longitude. However, visibilities are reported by Air Traffic Control in metres or kilometres. Pressure is quoted in millibars, but the hectopascal (hPa), which is equivalent to a millibar, appears on many charts issued by European Met. Centres.

Illustrations

The satellite pictures have two sources. The very high resolution pictures (those which have fine detail) were supplied by the University of Dundee Satellite Laboratory. Dundee produces the most magnificent results!

The low resolution pictures (those with lines of latitude and longitude added) were photographed from a video monitor using a WSR 524 bought from Feedback Instruments of Crowborough. This has the great advantage of automatic reception. An internal computer adds the geographic grid to each picture and retains it in memory until the next picture arrives.

The black-and-white photographs have been reproduced from colour slides, mostly taken from the cockpit of sailplanes and powered aircraft.

Air in motion

Where the power comes from

The atmosphere is an enormously energetic system. Practically all the power to run it comes from the sun whose radiation provides about 1.4 kW per square metre above the atmosphere. This energy does not all reach the surface of the earth; some of it is immediately reflected back into space. The proportion which is available to warm the surface varies across the globe. Equatorial regions, where the sun passes almost directly overhead, collect most of the energy. Polar regions, where the sun barely rises above the horizon, only receive a significant amount of energy during the summer months, and even then much of this is immediately reflected back into space by the cover of snow and ice.

The great contrast of temperature between tropical and polar regions produces an ever-changing flow of wind which redistributes the heat energy and produces the weather systems. The winds are a fundamental part of the weather process and this opening chapter describes something of the wind-producing mechanism.

First, a reminder of some basic features

Most people are familiar with the weather presentation on TV, or see little weather maps in the daily press. The section below may help to remind readers of some important features of the weather.

Pressure

Pressure decreases with height. The rate of change is shown by the lower (curved) line in fig. 1. The most commonly used unit is the millibar (mbar) but recently it has been replaced by the pascal which is only 1/100th of a mbar. Since this is inconveniently small some Met. services now label their charts in hectopascals (hPa) which are exactly the same as millibars.

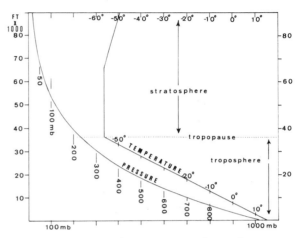

1. How pressure and temperature change between the surface and 90,000 feet

Table 1. Pressure units

1 Atmosphere	= 1013.25 millibars (mbar)
	= 1013.25 hectopascals (hPA)
	= 29.921 inches of mercury
	= 760.00 mm of mercury
	= 14.696 lbs per square inch
	= 1033.2 kg per square metre

At sea level a pressure change of 1 millibar occurs if you ascend 27 feet. Higher up this figure increases. At 3,000 feet each millibar equals a climb of 30 feet. At the height most jet aircraft fly (around 35,000 feet) you have to ascend about 90 feet to get the same pressure change.

The atmosphere

Figure 1 shows how temperature and pressure vary in the average atmosphere up to 90,000 feet. Figure 2 shows the temperature changes up to a height of 70 miles. These diagrams also show the different layers; each layer has a particular temperature structure.

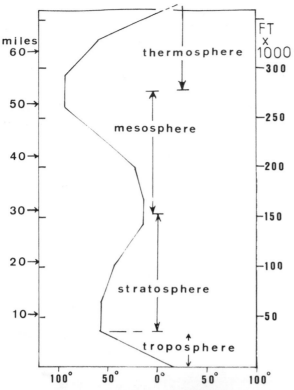

2. *Atmospheric layers and the temperature profile up to 70 miles*

The troposphere

Practically all the features which combine to produce 'weather' occur in the lowest layer of the atmosphere called the troposphere, where the average air temperature decreases about 2°C for every 1,000 feet of height. This layer is (on average) about 7 miles deep but it can vary between about 5 and 10 miles and nearly all the clouds and moisture are confined within it. Even the great storms which extend for a 1,000 miles or more are restricted to this relatively shallow layer of the atmosphere.

The stratosphere

Above the troposphere is a layer in which air temperature increases slowly with height. In this region the air is normally far too dry for any clouds to appear.

The tropopause

The boundary between the troposphere and the

stratosphere is called the tropopause. If you look out from a high-flying aircraft the tropopause can sometimes be seen as a level haze top, or a smooth flat cloud top, at about 36,000 feet. The level is not constant but undulates slowly up and down as major weather systems move beneath it.

The mesosphere and thermosphere

These are the two layers above the stratosphere. The temperature decreases in the mesosphere but increases with height in the thermosphere. At higher levels in the thermosphere the temperature becomes extremely high. However, there is so little air there that these high temperatures have no practical effect on the few astronauts who pass through them.

Isobars

These are lines of equal pressure. They are drawn on many weather maps to show up the different pressure systems such as anticyclones (highs), depressions (lows) and the associated valleys (called troughs) or their opposites (the ridges). Figure 3 illustrates these features.

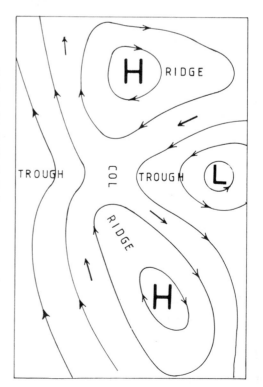

3. *Isobaric patterns*

Isobars and the wind

Over most of the globe, with the exception of regions near the equator, the wind blows almost parallel to the isobars. The strength of the wind depends on how close the isobars are spaced. Close spacing indicates a strong wind, wide spacing produces a light wind.

Buys Ballot's law

This states that in the Northern Hemisphere, if you stand with your back to the wind the lowest pressure lies on the left hand side. In the Southern Hemisphere, conditions are reversed and low pressure lies to the right when your back is to the wind.

Streamlines

In tropical regions, and particularly near the equator, the isobars no longer give an accurate indication of the wind. They are replaced by streamlines, which follow the direction of flow of the wind. Streamlines are not so easy to draw as isobars but they can show where the air flow is coming together (converging) or spreading apart (diverging). A comparison between an isobar chart and a streamline chart is shown in fig. 4. You may notice that the streamlines spiral out from the high and in towards the low and that several streamlines converge on the dotted line marking a frontal trough. By their nature isobars cannot meet in this way.

1940s when the availability of a large number of aircraft and balloon soundings of temperature made it possible to map the flow at high levels. To draw isobars one uses observations of pressure which have been corrected to give the value at some constant level, usually sea level. To draw contours one selects a convenient pressure and draws contours to show how the height varies. For example, the 1,000 mbar pressure surface is usually close to sea level. The actual height depends on the sea-level pressure and the air temperature. If one assumes a temperature of 15°C the relationship between contours and isobars is shown in Table 2 (below).

Table 2. Isobars and the 1,000 mbar contour height

mbar	metres	feet
960	− 344.5	− 1,130
970	− 253.7	− 843
980	− 174.0	− 571
990	− 84.8	− 278
1,000	0	0
1,010	84.0	275
1,020	167.1	548
1,030	249.5	818
1,040	331.0	1,086

The difference in height shows how the surface slopes. Table 2 also shows the kind of altimeter error one may expect when flying across the wind from high to low pressure. In the extreme case of a flight

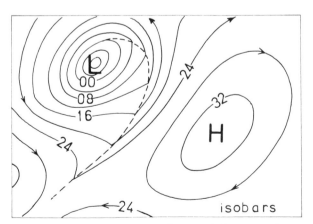
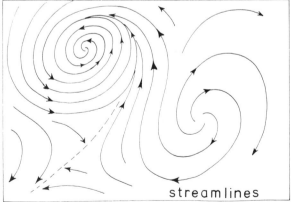

4. *Comparison between isobars and the streamline pattern*

Contour charts

The pattern of winds can also be represented by contour charts. These came into general use in the

from an anticyclone of 1,040 mbar to a deep low of 960 mbar the true height would be more than 2,200 feet **lower** than that indicated.

Using contours instead of isobars

The height of a pressure surface depends both on the sea-level pressure and the temperature of the air. The air temperature becomes a major factor at high levels. This can be observed when flying over the sea by comparing the height indicated by a pressure altimeter with the true height measured by a radio altimeter. For example, an aircraft flying at an indicated altitude of 30,000 feet (where the pressure is almost 300 mbar) would have a true altitude of about 32,000 feet in tropical regions (where the air below is warm). Flying towards the poles the radio altimeter would show a gradual decrease in height as the air below became colder. Eventually, the true altitude would be about 28,000 feet.

The change in height is due to the slope of the 300 mbar surface which the aircraft was following. Air expands when heated, so that a pressure surface rises when the air is warmed and sinks when the air is cooled.

Contours and the wind

Contour charts can be used just like the more familiar charts of isobars. The winds aloft blow almost parallel to the contour lines. Buys Ballot's law still holds. If you face down wind the lower contour heights lie to the left (in the northern hemisphere). The strength of the wind depends on the steepness of the slope. Close spacing of contours indicates a steep slope and consequently a strong wind. A wide spacing between contours shows a gentle slope which only produces a light wind.

Jet streams

Jet streams are bands of very strong winds which are usually found at levels between 30,000 and 40,000 feet, but which may appear at other heights. A jet stream is often more than a thousand miles long, one or two hundred miles wide and generally less than four miles deep. Most jet streams occur where there is a marked contrast in temperature, for example, where cold air from polar regions is brought close to much warmer air from the tropics.

Figure 5 shows an area close to America where jet streams often form, particularly in winter. The strong contrast in temperature makes the contours come very close together. As these contours draw closer, indicating a steepening of the slope, the wind speed increases.

The lower part of fig. 5 shows a cross section along the line A—B. The 300 mbar surface dips down towards the cold air and the jet stream occurs where the slope is greatest. Speeds of 100 knots are common, and speeds of more than 200 knots have been observed in the central core of some powerful jets. Once a jet has developed it can move off down wind and travel thousands of miles, snaking up towards polar regions and dipping down towards the edge of the tropics.

Figure 6 shows a 3-D diagram of a jet stream. The series of rings are isotachs which show how the wind speed increases towards the core of the jet. A jet is

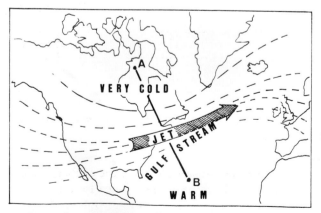
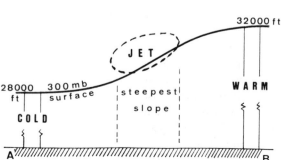

5. *Area where jet streams develop and the cross-section of the 300 mbar surface showing the steepening at a jet stream*

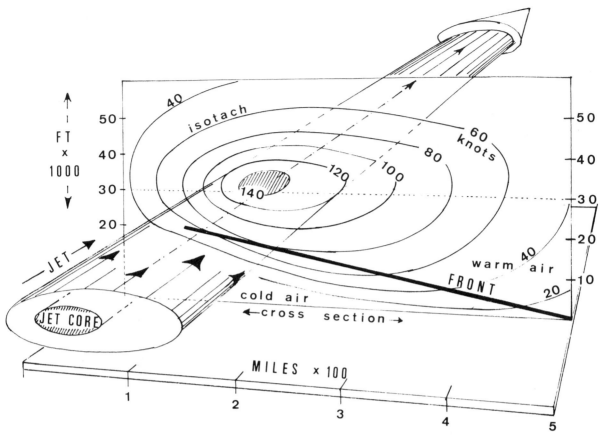

6. *Cross-section of a jet stream*

often associated with a 'front' (shown on this diagram by the sloping black line). Jets play an important part in the development of depressions. This is described in more detail in Chapter 2.

Geostrophic winds and how to measure them on a chart

The wind speed measured from a chart of isobars or contours is known as the geostrophic wind. It is usually a good approximation to the real wind but there are occasions (mentioned further on) when the actual wind differs significantly from the geostrophic wind. Meteorologists use a transparent scale (called a geostrophic scale) which is laid at right angles to the isobars. The wind speed is found from the spacing between isobars (see fig. 7).

Geostrophic scales are often printed near the corner of a weather map; they are more complicated than fig. 7 because they have to take account of variations

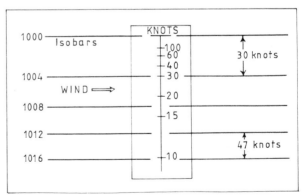

USING A GEOSTROPHIC SCALE FOR 4 MB ISOBARS

7. *Example of a simple geostrophic wind scale and its use*

of map scale and latitude. Each size of chart needs a different scale which is constructed for a particular interval between isobars. For example, in England isobars are usually drawn at intervals of 4 mbar. However in France and Germany one commonly sees a 5 mbar interval. This makes life difficult for anyone who travels to different countries. The problem can be avoided as follows.

Finding geostrophic winds without a scale

The process is illustrated in fig. 8.

(a) From the weather map measure off a distance of 5 degrees of latitude. This is equivalent to 300 nautical miles. Nearly all maps have lines of latitude marked on them. One can use a pair of dividers, mark a transparent piece of plastic or even mark the edge of a piece of paper.

(b) Lay the marked-out length at right angles to the isobars or contours in the area of interest and note the change in pressure or height between the ends.

(c) Multiply this by the factors in the table below to get the wind speed in knots.

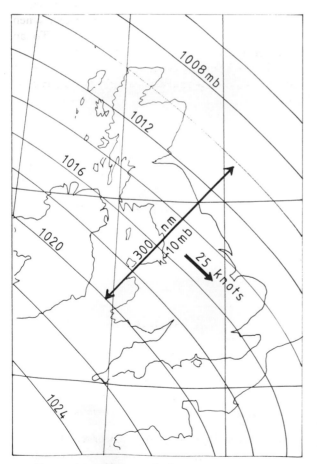

8. *Illustration of how to find a geostrophic wind without a scale*

Table 3. Factors to derive the geostrophic wind speed in knots from the change in pressure or height over a distance of 5 degrees latitude (300 nm)

Latitude	Multiplication factors	
	For isobars	For contours
degrees	(mbar)	(metres)
70	2.1	0.25
60	2.3	0.27
55	2.4	0.29
50	2.6	0.31
45	2.8	0.33
40	3.1	0.37
35	3.4	0.41
30	3.9	0.47
25	4.7	0.56

For example, if the surface chart (drawn in millibars) had a 10 mbar drop at latitude 52 degrees where the factor in the second column is about 2.5, then the wind speed would be $10 \times 2.5 = 25$ knots.

If instead of isobars the chart showed contours one would use the factors in the third column. Thus, if the contours showed a height change of 200 metres and the latitude was 45 degrees then (using the factor 0.33) the wind speed would be $200 \times 0.33 = 66$ knots.

Notice that as the latitude becomes less the multiplication factor becomes larger. Thus, a contour spacing which would give a wind speed of 50 knots at latitude 70 would give a wind speed of 94 knots at latitude 25. In low latitudes (towards the equator) where the factor becomes very large, the wind is seldom truly geostrophic and this kind of measurement becomes inaccurate.

How the wind changes near the ground

Geostrophic winds, whether they are measured from isobars or a contour chart, do not allow for any friction. The wind near the ground is slowed down by friction; the rougher the surface the greater the friction. It is generally assumed that the wind measured from the surface isobars will be about right at levels of 2,000 feet. Below that, the effect of surface friction slows the wind down. The wind direction then turns to blow across the isobars towards lower pressure.

There is no single relationship between the geostrophic wind and the surface wind; it depends on the surface and the time of day. For example, if the 2,000 foot wind was 270 degrees 20 knots then the surface wind would be about:

> 260 degrees 15 knots over a smooth sea
> 250 degrees 9 knots over land by day
> 240 degrees 7 knots over land at night
>   with some clouds
> 230 degrees 5 knots over land on a
>   cloudless night

Figure 9 illustrates how the wind speed may vary with height in the lower layers. Notice that the speed changes most rapidly close to the ground. The change of wind speed is very noticeable if you go across a high suspension bridge. When gales are blowing, high-sided vehicles such as furniture vans may have little trouble travelling along country roads with trees and hedges to reduce the wind speed; however, on a high suspension bridge the full strength of the gale may blow the vehicle over on to its side. Aircraft landing in strong winds need extra airspeed to compensate for the sudden drop in wind speed close to the ground. The lower part shows anemograph (wind recorder) traces for day and night. By day the trace shows hundreds of gusts and lulls over a few hours. At night the speed is lower and changes much less rapidly.

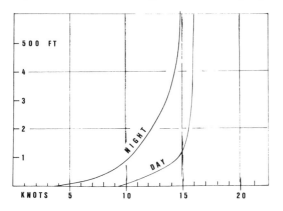

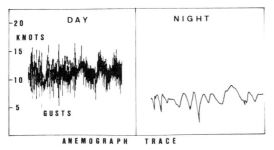

9. *The variation of wind speed with height and what anemograph traces look like by day and night*

How the geostrophic wind develops and why it can differ from the real wind

Many people do not need to bother with the details of the geostrophic wind. Unless you are particularly interested in the subject the rest of this chapter can be skipped.

Introducing the coriolis force

If the earth did not rotate on its axis there would be no geostrophic wind. The air would move directly towards a region of low pressure or flow down any slope shown by the contours. This would tend to prevent the deepening of nearly all depressions. However, the rotation of the earth deflects the moving air. This deflecting force is called the 'coriolis force'.

The upper part of fig. 10 represents the view looking down at the earth from directly above the north pole. The earth rotates in an anticlockwise direction as shown by the thick black arrow. When the air moves, the earth rotates under it. To an observer in space the air appears to be moving in a straight line but to an observer on the ground the air follows a path which curves to the right (shown by the pecked line).

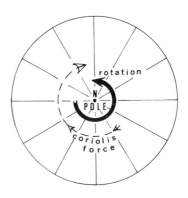

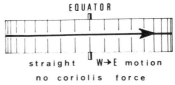

10. *Rotation of the earth and the coriolis force*

An observer high above the equator sees the earth's surface moving straight from west to east with no apparent rotation. No rotation means no coriolis force and no geostrophic wind on or near the equator.

To an observer above the south pole the earth's rotation is clockwise so the coriolis force produces a deflection towards the left in the southern hemisphere. The effect of the coriolis force is greatest near the poles; it varies as the SINE of the latitude to become zero at the equator.

How geostrophic balance is reached

The geostrophic wind is achieved when two forces are in balance; these are:

(a) the force of gravity pulling the air down a sloping pressure surface (or the pressure force when using isobars) and

(b) the coriolis force tending to turn the air to the right (in the northern hemisphere).

Figure 11 illustrates how the balance is reached. If a previously level pressure surface was suddenly tilted, the air would start to flow down the slope under the pull of gravity. As soon as it started to move the coriolis force would begin to deflect it. It would take several hours for this deflection to have its full effect. The effect is more rapid near the poles than at low latitudes. Eventually the path of the air is turned through a right angle so that it flows parallel to the slope (or to the isobars). The coriolis force still tends to deflect the wind but if the path of the air were turned any further it would start to flow uphill. Instead, a balance is reached when the pull of gravity is exactly balanced by the coriolis force. Now we have a geostrophic wind.

To summarise: the geostrophic wind blows parallel to isobars or contours with low pressure (or height) to the left in the northern hemisphere. The wind speed depends both on the gradient and the latitude. When the gradient is steep the wind is strong; for a given gradient the speed increases as the latitude decreases. The relationship breaks down near the equator where the coriolis force vanishes.

When the wind is not geostrophic

If the wind was always perfectly geostrophic most of the familiar weather systems would never develop. In fact, the geostrophic balance is easily upset and when this happens the air currents can converge and produce areas of ascent where cloud and rain develop.

The gradient wind

The gradient wind is a modification of the geostrophic wind to take account of curved flow. When the isobars/contours are curved a third force is introduced; this is the centrifugal force, familiar to all who drive round corners rapidly. The effect on the wind speed depends on whether the curve is cyclonic (round a low or trough) or anticyclonic (round a high or ridge). Figure 12 shows the two possibilities in the northern hemisphere where cyclonic curves are those which turn to the left.

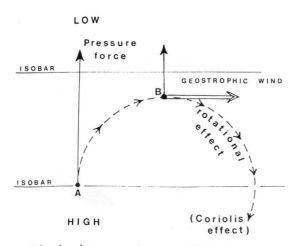

11. *The development of geostrophic balance*

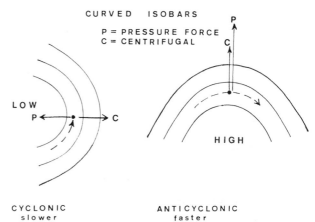

12. *Illustrating the effect of curved isobars on wind speed*

With cyclonic curvature the pressure force (P) acts inwards but the centrifugal force (C) acts outwards. As a result the effect of the pressure force is reduced and so the wind speed is reduced. With a very small radius of curvature (about 100 miles) the wind may be reduced to half the speed for straight isobars.

When the curvature is anticyclonic both the pressure force and the centrifugal force act in the same direction (outwards). This increases the effect of the pressure force and so the wind becomes much stronger. The wind flowing round a marked ridge may be increased to nearly twice the speed for straight isobars. This is the upper limit, beyond it the flow ceases to be steady.

Looked at another way, one cannot have closely-spaced isobars near the centre of a high because the wind speed would exceed the limit for an anticyclonic curve and the air would slide outwards instead of following the bend. This would bring down the pressure in the high and reduce the gradient to permitted limits. It is different round a low; if the air slides outwards it will reduce the pressure in the centre and increase the gradient.

When the lines are not parallel

If the isobars/contours are not parallel the geostrophic balance is upset for a different reason which may be summarised as momentum (or the lack of it, inertia). At sea level the mass of air above every square mile weighs about 26 million tons. This mass takes time to respond to a change in the pressure gradient or an alteration in the slope of contours.

Figure 13 shows two possibilities. In (a) the lines fan out; the gradient is decreasing down wind. The air enters from the left with considerable momentum and the speed is soon too high for the gradient. The pull to the left becomes less as the lines fan out but the coriolis pull to the right remains the same until the air slows down. For a time the coriolis force is the stronger and the air turns across the isobars towards the right. The faster the air is travelling when it reaches this region the more noticeable is this right-hand deflection. It quite often amounts to 20 or 30 degrees at the levels where long distance aircraft fly.

The opposite effect occurs when the lines converge down wind. Then the inertia prevents the air accelerating fast enough. Since the air is moving too slowly the coriolis force is too weak. The strengthening pressure force then takes the air across the isobars towards the low, as shown by the pecked line.

Rapid pressure changes

A similar effect can be noticed when there is a rapid fall of pressure. Once again the wind fails to respond to the new conditions and the air starts to flow across the isobars towards the region where the pressure is falling most. The component across the isobars is called the 'isallobaric wind'. Its effect is most marked when the original wind speed is fairly light. Then the rapid fall of pressure (ahead of an advancing low or trough) can make the wind direction swing 45 degrees or more and turn inwards towards the trough.

Friction

Frictional drag slows down the air near the ground. This upsets the geostrophic balance. The slower the air moves the less effective is the coriolis force. The result is that the slower air is deflected towards the low pressure side, shown in fig. 14.

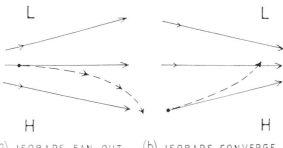

(a) ISOBARS FAN OUT   (b) ISOBARS CONVERGE
    AIR CURVES RIGHT   AIR CURVES LEFT

13. *The effect of isobars which fan out or converge*

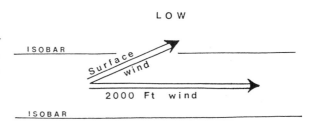

14. *The deflection of surface winds due to frictional drag*

Convergence and divergence

Winds which differ from the geostrophic are termed 'ageostrophic'. When they occur the flow of air develops regions of 'convergence' where the streamlines (but not the isobars) come together, or 'divergence' where the streamlines move apart. Convergence is rather like a line of traffic slowing down as it approaches a road junction or an obstruction. After a time all the vehicles are nose-to-tail and the load on the road is increased. Divergence is the opposite effect; once past the obstruction cars speed up and the traffic becomes well spaced out.

If the air started to converge at all levels the excess would cause a rise of pressure at the surface. When traffic approaches a hold up some drivers may turn off down a side street rather than join the queue. In the atmosphere convergence at one level often results in some of the air moving to another level where the air is diverging.

When air flow converges at low levels but diverges higher up some of the air is forced to ascend into the divergent region. Air cools as it rises and cooling eventually leads to condensation of moisture to form clouds and rain. This can be the start of a major bad weather system. The process, which is often initiated by an acceleration in the fast-moving flow at levels some 4 to 8 miles up, is a major factor in the formation of large depressions and anticyclones. The process is described in Chapter Two.

2

The development of depressions and anticyclones

Depressions are responsible for much of the bad weather which occurs; this section is an account of some of the reasons why areas of low and high pressure develop. Two important factors are:

(a) a large supply of heat to provide the energy to transport huge masses of air, and

(b) regions where the wind is no longer geostrophic so that the air converges in one place and diverges at another.

Making a little depression

This process can start from calm conditions both at the surface and up aloft. It works best over land when there is strong sunshine. Figure 15 illustrates this.

(a) Sunshine warms the ground and heat is then transferred to the air above. If conditions are suitable this heat is carried upwards by convective currents so that a broad column of air (say 50 miles in diameter) becomes warmer than its surroundings.

(b) This warm column expands and pushes up the pressure surfaces to form a dome in the contours aloft.

(c) The air forming the dome then starts to move directly downhill under the pull of gravity. Given a quarter of a day the coriolis force would deflect the flow until it became parallel to the slope, but to begin with the airflow is not geostrophic and diverges in all directions.

(d) This divergence aloft reduces the total amount of air below, so the surface pressure starts to fall. The little low has started to form. This is how 'heat lows' form over land on sunny days. It is not the most efficient way of forming a new low because the energy supply is cut off every night but it is effective in tropical regions.

(e) Once there is a surface low the air all round begins to move in to fill it up. Again, the coriolis force takes time to deflect this inward flow and low-level convergence acts to limit the pressure fall.

The process can be improved when the air is moist enough for the rising air to condense and form clouds. Condensation provides an additional supply of energy by releasing latent heat. A wide column of convection cloud can become a sort of chimney. Once the low has formed, inflow at the bottom provides a steady supply of fresh moist air. Ascent releases latent heat which warms the air rising up the 'chimney'. The air eventually spreads out sideways from the top. This is one example of the kind of 'feedback' which often enables an initially weak process to become much more effective with time.

Some 'polar lows' are thought to develop in this

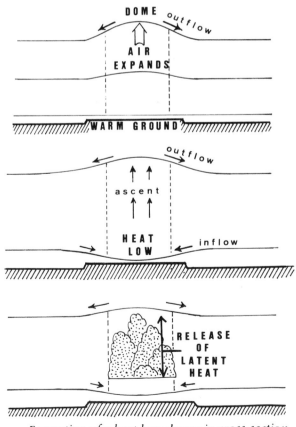

15. Formation of a heat low shown in cross-section

way when very cold air moves out over the relatively warm ocean.

Making a large depression

The process described above is not powerful enough to generate a really big depression though some of the features assist in the formation of tropical depressions.

The depression-making process is much more efficient if it is started off by a strongly divergent flow at high levels.

Chapter One ended with a paragraph on convergence and divergence. This kind of flow is closely connected with another feature—'vorticity'.

Introducing the idea of vorticity

Vorticity is the term used to describe the amount of spin. It is made up of two parts, shear and rotation, illustrated in fig. 16. In (a) there is straight flow with

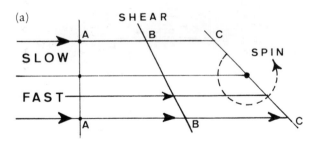

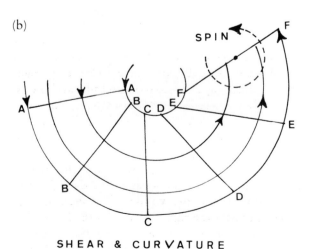

SHEAR & CURVATURE

16. *Vorticity due to (a) shear and (b) shear and rotation*

slower winds on the left of the flow and stronger winds on the right. Suppose an aircraft left a very long-lasting condensation trail along the line A–A. As the trail moved along with the wind it would slowly rotate, first to B–B and later to C–C. This reveals the amount of spin (vorticity) produced just by the shear of wind.

The spin is greater if the air also flows round a curve as in (b). In this case the starting line A–A is eventually twisted round to lie along F–F.

Vorticity and divergence

It is an everyday experience to see the vortex which forms when a basin of water empties down the plug-hole. There is always some slight vorticity in baths or basins and the asymmetric flow towards the plug introduces more. When some water starts to drain out the rest begins to converge upon the exit. The spin, which was originally too slight to be noticed, is increased where the water converges. Soon it becomes so great that a visible vortex forms where the water spins round very rapidly before disappearing down the plug-hole.

The effect is reversed if the water empties into a large container where it can spread out (diverge). The spin then rapidly disappears.

In the atmosphere the most striking examples of vorticity are tornadoes and waterspouts. These are found where the air converges towards the base of a thunder cloud into which the air is being drawn very rapidly. Little dust devils are produced in the same way on hot sunny days when there is a vigorous ascent of air. The difference is that dust devils are much smaller and can develop without a cloud above them.

Large depressions which extend for hundreds of miles are sometimes called vortices. They may not seem to have much spin when seen on a single satellite picture but a time-lapse series often shows the cloud swirling round the centre.

Vorticity can be calculated but it is a tedious exercise best left to computers. All that most of us need to know is that vorticity is increased when air flows through a trough and decreased when it flows round a ridge. This implies that air will (nearly always) be converging as it goes through the trough and diverging when it rounds a ridge.

Ascent of air where upper and lower troughs are out of phase

When charts for the flow aloft are superimposed on surface charts one frequently finds that the surface trough lies down wind of the upper trough. This is shown in fig. 17. The surface flow is shown by full lines and the upper flow by pecked lines. The upper trough lies to the left (upwind) of the surface trough. The vorticity is at a maximum along the trough lines (shown on the lower part of the diagram as 'VORT MAX').

As the air approaches this region there is convergence; when it moves on into decreasing vorticity there is divergence. When the upper and lower patterns are superimposed we can see that the convergence at low level lies under the divergence at high level. In this region air is forced to ascend from the low level trough.

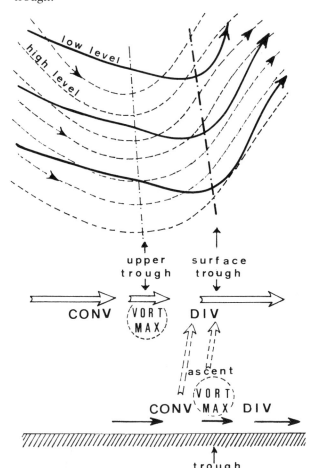

17. *Ascent of air where surface trough precedes an upper trough. Plan view and cross-section of ascent*

Jet streams and the formation of lows

To get strong divergence it is usually necessary to have a fast-moving current of air aloft. In jet streams, where there is a band of air travelling at speeds of 100 knots or more, there is strong wind shear. When this is combined with a curved flow the air experiences big changes of vorticity; it speeds up and diverges in one region and slows down and converges in another. The stronger the winds aloft, the more marked can be this divergence.

Figure 18 shows a straight section where the contours are close together in the middle indicating a jet stream. As the air approaches the jet it starts to move faster. There is divergence and an inflow along the dotted line from the region marked 'IN' near the jet entrance. At the other end (the jet exit) the air starts to slow down in the region of convergence; the pecked line shows that some air moves across the contours to the region marked 'OUT'.

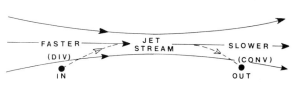

18. *Divergence and convergence in a straight jet stream*

This type of flow has been examined by tracking constant pressure balloons through a jet. The balloons followed tracks similar to the pecked lines. At the jet exit some of the balloons curved so far to the right that they ended up going round in right-handed circles. It is believed that one of the early, unsuccessful attempts to cross the Atlantic by balloon failed because the balloon left the jet exit and got stuck in the right-handed swirl.

Figure 19 shows a jet stream with a trough upstream on the left and a ridge downstream on the right. The curvature adds to the vorticity and increases the divergence near the right entrance of the jet. With a pattern like this it is common to find that a surface depression forms under the region marked 'DIV' while an anticyclone develops in the region below the 'CONV'.

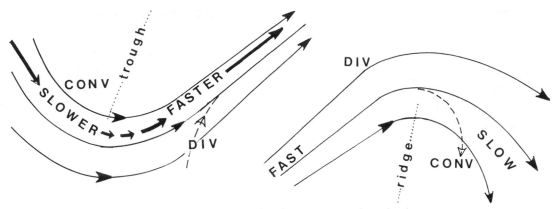

19. *Divergence and convergence when jet stream lies between trough and ridge*

Figure 20 is a vertical cross section illustrating this process. In (a) the ascent of air into the region of upper divergence produces a fall of pressure and low-level convergence near the surface. In (b) the slowing down of the jet in the region of upper convergence results in a rise of pressure at the surface below and descent of air.

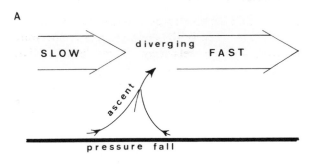

20. *Vertical motion and surface pressure change beneath regions of high-level divergence and convergence*

Effect of low-level convergence on the development of a front

When pressure begins to fall at the surface the low-level air starts to converge towards that region from all sides. Colder air from polar regions is brought closer to warmer air from the sub-tropics. This increases the contrast between the cold and warm air masses. The convergence is maintained by ascent of air from low to high levels; it eventually draws warm and cold air masses so close together that a well-defined boundary forms between them. This boundary is termed a 'front'. The process is illustrated in fig. 21 (a) and (b). Figure 22 shows the cross section of a front. The warm air, being less dense than the cold, lies above it and the frontal surface forms a sloping boundary at an angle of roughly 1 : 100. (The slope is not constant.)

Feedback from low-level convergence

The low-level convergence not only intensifies the temperature contrast between warm and cold air masses, it also steepens the slope of the pressure surfaces aloft and consequently strengthens the jet stream. Here is an example of positive feedback: the jet started low-level convergence, the low-level convergence increased the temperature contrast, and this in turn strengthened the jet.

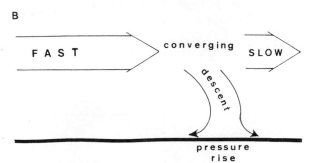

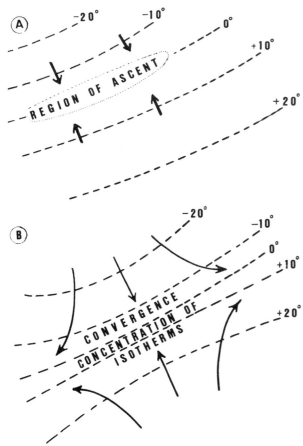

21. *How convergence can produce fronts by concentrating isotherms*

Adding more energy to the system

When the low-level air starts to ascend, a new source of energy becomes available. The rising air cools and the water vapour starts to condense to form clouds. Condensation releases latent heat. If the rising air is warm it can contain a large amount of water vapour. Each kilogramme of tropical air may hold some 20 grammes of water vapour at sea level. By the time this air has risen 3 miles the cooling will have made half of the vapour condense as clouds, releasing nearly 600 calories per gramme. This adds a lot of extra energy to the system.

Lows become much deeper when the rising air carries a lot of moisture in it. This is one reason why depressions which draw in warm, moist air from the tropics can become much larger and more intense than those which develop in dry desert regions, or in cold arctic air where there is very little moisture.

Life history of a frontal depression

A large number of depressions develop on an old front where there is already a marked temperature contrast over a relatively short distance. The sequence of events is shown in fig. 23.

On the left-hand side the sequence from (a) to (e) shows how the pattern changes at the surface and up

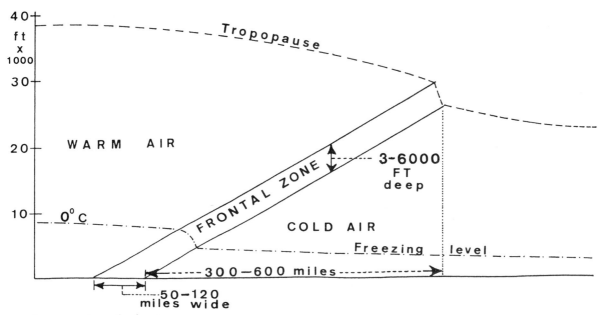

22. *Cross-section of a front*

aloft. The right-hand side shows the same sequence viewed from a satellite.

The sequence is drawn for the northern hemisphere. Developments in the southern hemisphere are a mirror-image of this. Isobars are shown by full lines; pecked lines show the 300 mbar contours which represent the flow at about 30,000 feet. The band of strongest winds is shaded in and marked 'JET'. These two levels have been combined to illustrate the interaction between high- and low-level flow in the development of a depression.

(a) shows the initial stage when the front is quasi-stationary (hardly moving). The surface isobars are very widely spaced indicating light winds. Up aloft there are strong winds blowing almost parallel to the front. Apart from a slight curve at the left the upper contours run almost straight at this stage. The strongest part, the jet, runs to the north of the surface front. A small dotted circle near the right-hand entrance to the jet marks the region where pressure begins to fall.

(b) is the next stage: a depression has formed and deepened enough for the isobars to show a closed circulation of air at low level round the centre. This circulation has begun to twist the line of the front. At this stage the system is called a 'frontal wave' or a 'cold front wave'.

On the western side of the new low the flow has become northerly; this takes colder air southwards. The upper flow responds to this by developing a more marked trough west of the surface centre. The increased curvature round this upper trough increases the upper divergence where the winds accelerate away from it. This in turn increases the pressure fall at the surface (another example of positive feedback).

At (c) the surface low has become fairly deep and the front has been twisted into an inverted 'V' shape. The inside of the V is called the 'warm sector' because it contains the warmest air. On the eastern side, where the cold air is being replaced by warm air, the front is called a 'warm front' and has been identified by rounded blobs. On the western side, where warm air is being replaced by cold, it is called the 'cold front' (identified on charts by triangular spikes).

At high levels the flow has become much more curved, with a well-marked 'upper trough' in the contours and an S-bend in the jet stream. This is usually the stage when the depression deepens most rapidly.

(d) shows that part of the warm front has been overtaken by the following cold front to form an 'occlusion'. There is still a large warm sector but it is separated from the centre of the low. This is an important step because a significant part of the energy needed to form this low came from the latent heat released when the warm air was lifted to form cloud.

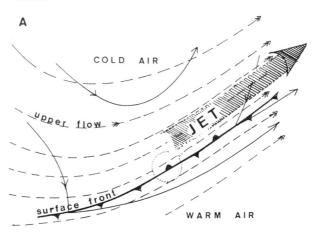

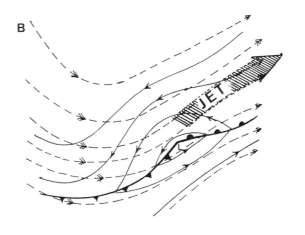

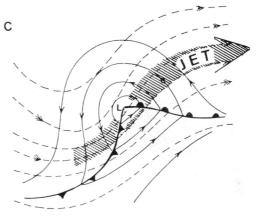

Another important change has taken place aloft. The axis of the jet stream has been transferred and now lies across the tip of the warm sector. The upper trough to the west has become so deep that a small circulation has formed south-west of the surface low. In the early stages the circulation was confined to a shallow layer near the surface. As the depression deepened the circulation extended higher up. At this stage it has reached 30,000 feet but later it may extend into the lower reaches of the stratosphere.

These two stages, occlusion (leading to separation of warm air from the centre of the low), and the growth of a cyclonic circulation up to high levels, mark the end of the most active period. From now on the main activity is transferred to the outer regions where there is still a supply of warm air and the jet stream is active.

(e) is the final stage: a long occlusion is wrapped round the centre which is filled with cold air. There is no warm air for hundreds of miles. The cyclonic circulation is firmly established up to the base of the stratosphere and there is now very little to displace the centre. The jet stream is far away but still powerful. A little to the south of the jet the cold front has come to a halt and a new wave is beginning to develop. This is the start of another cycle.

A satellite view of the developing low

The right-hand side of fig. 23 shows how the development appears on satellite pictures. The fronts are marked by thicker black lines. Lower clouds are shown with single hatching, thicker clouds whose tops reach much higher are double hatched.

(a) The quasi-stationary front is marked by a long but fairly narrow band of cloud. In one section the band is thickened by a small bulge on the northern side. This is the first sign of wave development and it often appears on satellite pictures before the surface charts show any sign of a new low.

(b) Now the cloud over the bulge has grown deeper; the cross-hatched region shows where the main ascent of air has taken place. A wedge of high cloud is drawn out into the jet stream to form a finger pointing almost parallel to the front. On infra-red pictures, which emphasise any temperature contrasts, this cloud appears as a bright, white streak showing that it is very cold (and therefore very high).

(c) This is the stage of maximum deepening when the warm air is rising over a wide area to form a shield of cirrostratus ahead of the depression. On high-resolution pictures one may see that the cirrus edge consists of filaments which fan outwards suggesting that the flow aloft is diverging. When such fanning out appears it is a reliable sign of marked deepening at the surface.

A

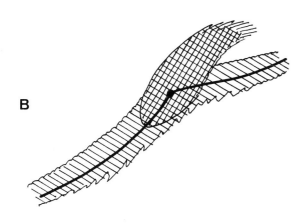

B

C

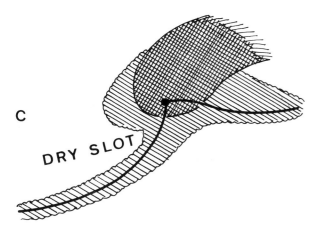

DRY SLOT

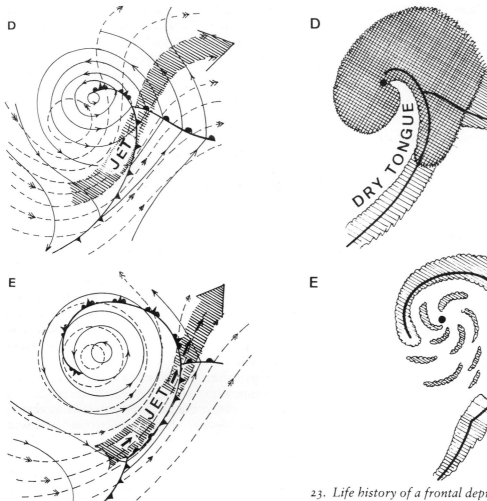

23. *Life history of a frontal depression shown by charts and satellite pictures*

The western flank of the cloud mass has a small indentation. This is called the 'dry slot'. It marks the incursion of very dry air which has descended from high levels, sometimes from the stratosphere. Its appearance is the first indication of the separation which will develop between the central part of the low and the warm, moist air to the south.

(d) This is the appearance when the depression is near its deepest stage. Occlusion has taken place and the warm air is being lifted clear of the surface. A long, clear region known as the 'dry tongue' marks the spread of dry air curving round the southern side of the low. There is a large mass of thick cloud but much of it is wrapped round the centre instead of being blown out ahead. This wrap-around indicates that the circulation is no longer confined to the lower levels but now extends up to cirrus levels as well.

(e) In the final stage there is little left of the warm air near the centre. The occlusion is marked by a thinning band of cloud curving round the centre. Closer in, where the air is both cold and unstable, there are arcs of cloud. These are curving lines of Cu and Cb which rotate round the centre. Some depressions are marked by almost continuous lines curving into the centre like a spiral nebula on an astronomical photograph.

Explosive deepening: 'bombs'

When one large depression becomes slow moving and starts to fill up, secondary lows often form on its perimeter. These secondaries usually start as waves on the cold front. The waves then deepen, grow larger and become major lows. The sequence may be repeated a number of times and the whole series is called a 'family of depressions'.

Occasionally one member of this family starts to deepen with exceptional rapidity. The Americans have called them 'bombs'. The central pressure of a bomb falls at more than 1 mbar per hour for at least 12 hours. A shallow low starting off at about 1,010 mbar would be down to 960 mbar within 2 days.

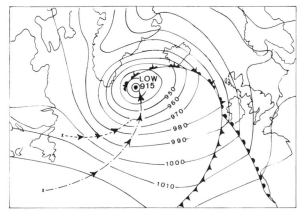

24. *The exceptionally intense depression of December 1986*

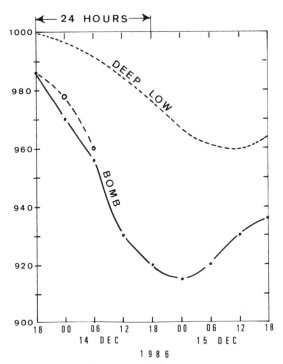

25. *How central pressure fell in a normal deep low and in a 'bomb'*

The deepest Atlantic low recorded in recent years started as quite an ordinary wave of 986 mbar south of Newfoundland at 1800 hrs on 13 December 1986. Thirty hours later it had deepened by at least 70 mbar and was centred between Iceland and the southern tip of Greenland. The exact pressure at the centre is not known; the UK analysts drew it as 916 mbar but the West Germans suggested 912 or 913 mbar. Figure 24 shows the storm at its most intense when the circulation covered a large part of the North Atlantic. Figure 25 shows how the central pressure dropped and compares the bomb with a more normal deep depression.

The atmosphere needs to be wound up fairly tightly before a new low turns into a bomb. The first of a family of lows does not deepen so dramatically. It merely sets the stage for its much fiercer offspring. In the December 1986 example the wave developed over the Gulf Stream where it had a supply of warm, moist air below and a jet stream above. There was an extra factor on this occasion—another secondary low following a converging track to the north. The tracks met, the lows merged and then came the period of exceptional deepening. For a time the central pressure fell at more than 4 mbar per hour. (*Note:* barographs show that in any one place the pressure may change even more rapidly; this is because the instrument responds both to the central deepening and the motion of the low.)

Fortunately most bombs reach their peak intensity out over the oceans, far away from the centres of population. When they do cross land the result can be disastrous. During the afternoon of 15 October 1987, two secondary lows merged over the Bay of Biscay. The combination deepened rapidly to about 958 mbar and wound up into an intense depression which moved across central England during the early hours of 16 October. In a sector to the south and east of the centre, surface winds gusted to 100 knots or more over Brittany and the Cherbourg Peninsula and reached 90 knots near the south coast of England. There was a lot of structural damage and millions of trees were blown down.

The track of depressions

The flow aloft is usually the most important factor controlling the movement of a depression. Figure 26 shows the kind of track followed by the system

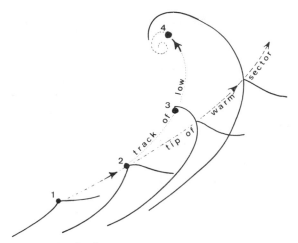

26. *Track of a depression and movement of fronts*

By stage (c) the upper trough has deepened so much that a closed centre has formed in the contours aloft. This swings the surface low round on to a northerly track.

The final stage (d) has an almost solid circulation extending from the surface up to the base of the stratosphere. There is little to steer the old low and the centre becomes very slow moving, often describing tiny circles like the wobbles of a top running down.

Old lows like this do start to move again when the system becomes asymmetrical (with light winds on one side and strong winds on another). The centre then moves in the direction of the strongest winds. This is shown in (e) where the strongest winds are on the south side (blowing from a westerly direction). The centre then moves east.

An example of a deep low from peak activity to decline

Figure 28 (a–e) illustrates a series of five infra-red pictures from NOAA–9 and NOAA–10 received between 31 January and 1 February 1988. Cold, high clouds appear very white, lower clouds are grey, the sea (being relatively warm) appears black. Lines of latitude and longitude are marked every 5 degrees.

Picture (a) on the morning of 31 January shows a

illustrated in fig. 23. The way that the track is influenced by the flow aloft is shown in fig. 27. Surface isobars are full lines and the upper flow is represented by pecked lines.

In (a) the surface low is only a shallow feature and the upper flow is a straight west to east current. At this stage the new low travels east.

In (b) the deepening of the surface low has begun to twist the upper flow bringing the high level winds back to a south-westerly direction and steering the low north-east.

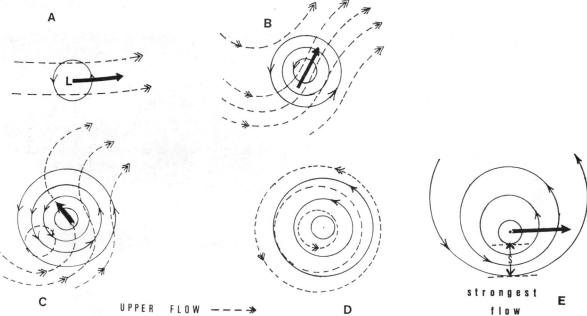

UPPER FLOW - - - ➔

27. *How upper flow controls track of surface depression*

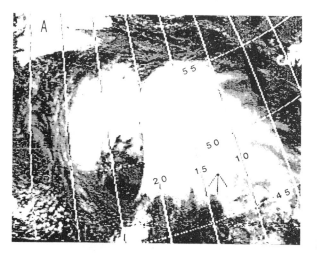

similar stage to fig. 23 (d) when the occlusion was already well developed. Pictures (b) and (c), for the afternoon and evening, show the swirl of cloud winding up round the centre of the depression. Notice that the highest cloud (shown white) begins to disappear where the occlusion is being wrapped around the low. Pictures (d) and (e), taken in the early morning and mid-afternoon of 1 February, illustrate the decaying stages. The curved band of frontal cloud has moved well out from the centre which is marked by curved segments of Cb clouds.

28. *(a) to (e) Satellite views of a depression from peak activity to decay*

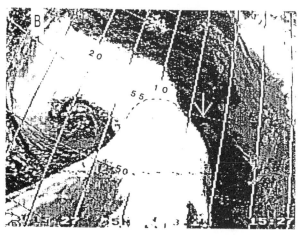

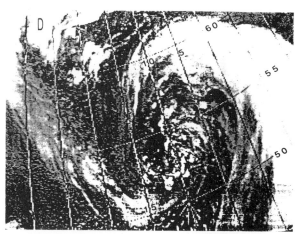

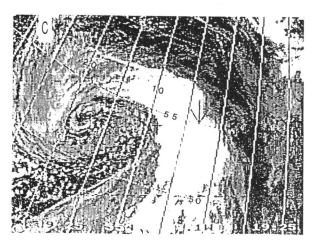

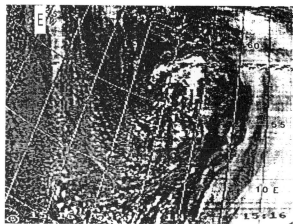

3

Fronts

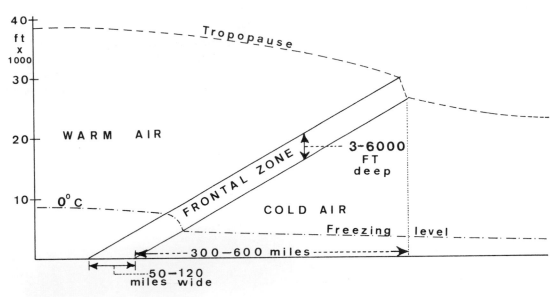

29. *Cross-section of a frontal zone*

A front is the boundary between two air masses of different density. The frontal surface usually has a slope of about 1 : 100 with the warmer (less dense) air lying above the colder air.

Figure 29 shows a cross-section of a front. The tropopause (which marks the base of the stratosphere) is shown by a pecked line with a kink where the front extends up to it. On some occasions the

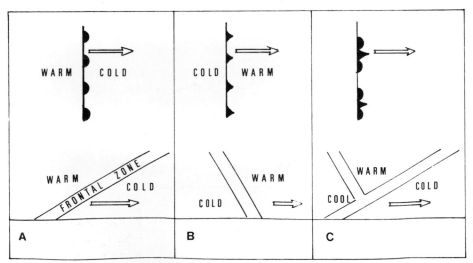

30. *(a), (b) and (c) How fronts are marked on charts and what their cross-sections look like*

tropopause may be broken or folded over at that point. The freezing level is also marked to show how the level changes across the frontal zone.

Figure 30 shows how fronts are marked on surface charts and what this indicates about their vertical structure:

(a) is a warm front—the surface position is shown by a line with semi-circular blobs on it; on hand-drawn charts it is coloured red. It is called a warm front because cold air is being replaced by warm air.

(b) is a cold front marked on the surface chart by triangular spikes or coloured blue. The warm air is being replaced by cold air.

(c) shows an occlusion. Here the cold front has overtaken the warm front. The cross-section shows that the warm air has been lifted above the surface.

An occlusion can have two structures depending on whether the overtaking cold air is colder or warmer than the air which was originally ahead of the warm front. This is shown in fig. 31.

In (a) the preceding air is warmer than the following air so the occlusion is called a 'cold occlusion'. The cross-section shows that the cold occlusion is more like a cold front at low levels. It has undercut the warm front aloft.

In (b) the coldest air lies ahead (to the right) of the occlusion. Since it is being replaced by slightly warmer air entering from the left the front is called a 'warm occlusion'; it has some of the character of a warm front.

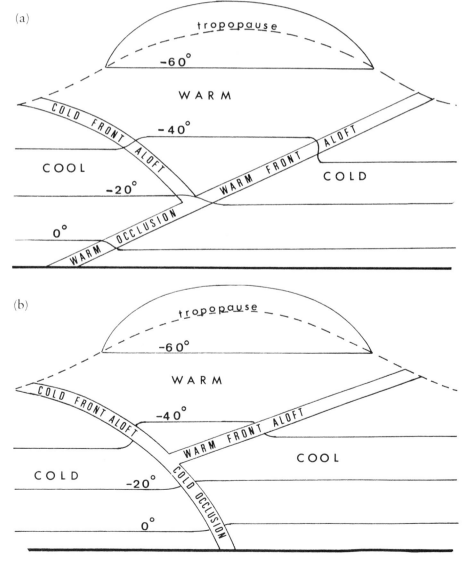

31. (a) and (b) The occlusion: two types of structure in cross-section

Plan and cross-section of fronts

Figure 32 shows a plan view of a warm sector depression as it appears on a surface chart. The lines A–A, B–B and C–C mark the position of the vertical cross-sections through the fronts. These are idealised sections. In reality the cloud and weather may vary considerably along the fronts; it depends chiefly on whether the air flows up the frontal surface or not.

Figure 33 illustrates the general distribution of rain (shown hatched) and the main cloud boundary (shown by the outer line). In this example most of the rain appears near the tip of the warm sector and ahead of the warm front. The cold front has a narrower rain

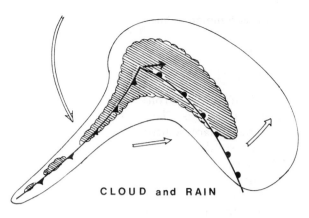

CLOUD and RAIN

33. *Cloud and rain in a warm sector depression*

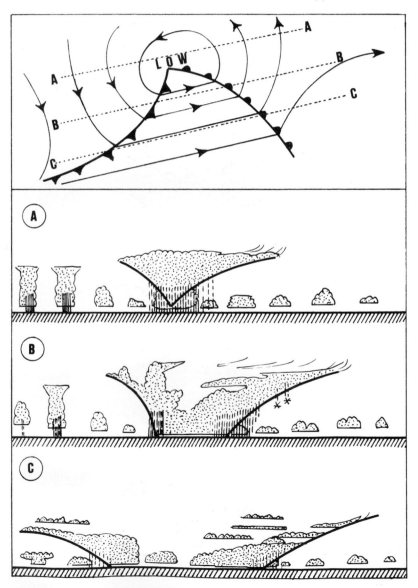

32. *A warm sector depression with three cross-sections through it marked A, B and C*

belt. It is common to find that when the warm front has a wide rain belt the cold front is less active, and vice versa.

Anafronts and katafronts

Cloud and rain depend on the way the air moves over the frontal surfaces. The air does not always ascend above this surface; sometimes there is a slow, sinking motion instead.

If air rises up the frontal surface it is called an 'anafront'. When the air is forced to rise, the cooling causes cloud and rain to develop and the front is an active feature. However, if there is general descent of air the front is called a 'katafront'. This is shown in fig. 34. Descending air warms and becomes relatively dry, thus dispersing much of the cloud and eliminating most if not all of the rain.

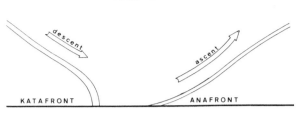

34. *Katafront and anafront*

Figure 35 (a) shows a cross-section of an ana warm front. The air is ascending from left to right and producing deep layers of cloud and rain.

To summarise: anafronts are active because the air ascends; katafronts are weak because the air descends.

Figure 35 (b) shows a kata warm front where the descent of air aloft has dispersed all the upper cloud and left only low cloud.

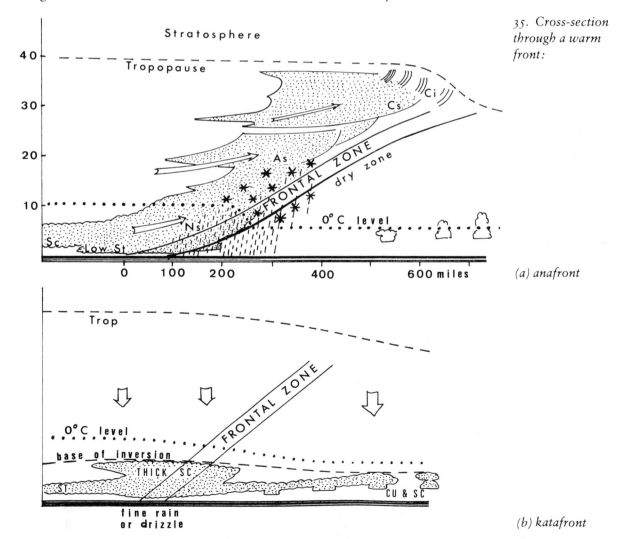

35. *Cross-section through a warm front:*

(a) anafront

(b) katafront

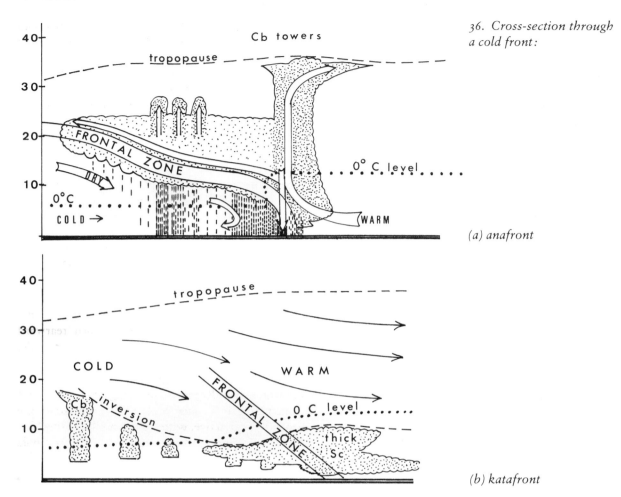

36. *Cross-section through a cold front:*

(a) anafront

(b) katafront

Figure 36 (a) shows the ana cold front. The frontal surface is steeper than at a warm front. The strong ascent above the front has produced cumulonimbus as well as layered cloud.

Figure 36 (b) shows a kata cold front where the air descends above the frontal surface, dispersing most of the cloud and rain.

These diagrams illustrate the importance of the flow over the frontal surface. Unless this flow is directed upwards the front remains a weak feature as regards cloud and weather. There is still a temperature contrast between the two air masses even if the front is not marked by a rain belt.

The conveyor belt

A frontal system does not move as a solid block of air. It can have quite a complicated set of air currents flowing through and round the system. In recent years the term 'conveyor belt' has been used to describe these currents. A conveyor belt is a long and fairly wide band which carries vast quantities of air through the low pressure system, often travelling a thousand miles or more and sometimes ascending many thousands of feet during the journey.

There are basically two main conveyor belts, the major one is known as the 'warm conveyor belt' and this brings with it most of the warm moist air which eventually produces thick cloud and extensive rain.

The warm conveyor belt

The warm conveyor belt is a band of air which generally starts off between latitude 25 and latitude 40. Initially it is some 3,000 feet in depth and 100–700 miles in width. The air moves ahead of a cold front and often flows nearly parallel to it for many hundreds of miles. Figure 37 (a) shows a plan view of a warm conveyor belt running parallel to the cold front and then ascending over the warm front; (b) shows a three-dimensional view. In this sketch the conveyor belt is shown by pecked lines. The surface position of the fronts are marked with the usual symbols and the sloping surface shown by solid lines. The position of the jet-stream core above the frontal surface is drawn with dotted lines and a chain of arrows.

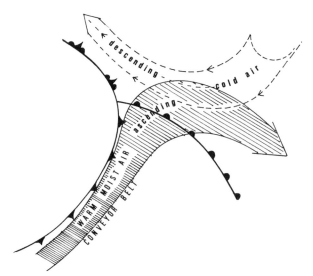

37. *Warm and cold conveyor belts; (a) plan view with fronts*

Notice that in this diagram the conveyor belt does not rise over the cold frontal surface except near the tip of the frontal wave. As a result most of the thick cloud and rain is concentrated near the warm front while the cold front brings very little rain. It is common to find that active warm fronts are followed by relatively weak cold fronts, or vice versa. An important factor is the alignment of the conveyor belt.

Figure 38 (a) and (b) overleaf show two kinds of rainfall areas associated with a conveyor belt rising over a warm front. Light rain is shown by single hatching, heavier rain is marked by double hatching. Notice that in each case rainfall starts some distance before the conveyor belt actually reaches the surface front. This is because the top of the belt starts rising much earlier and produces cloud thick enough for rain to develop in the warm sector.

Figure 39 (a) illustrates a warm conveyor belt rising up the following cold front. This is called 'rearward sloping' because it slopes backwards relative to the advancing cold front. The main areas of rainfall have been shaded in. Figures 39 (b) is a 3-D sketch of this rearward slope. Rearwards-sloping flow up the cold front makes it an ana cold front. Such fronts are generally very active, with cumulonimbus rising high above the front and heavy rain with squally winds below.

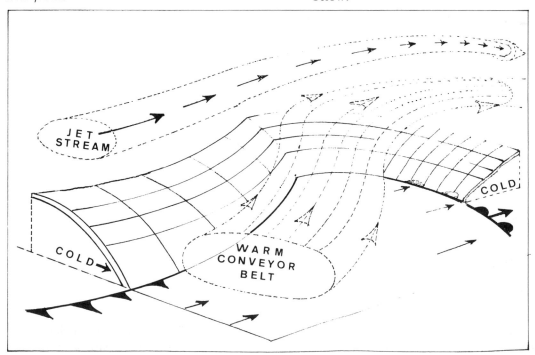

(b) 3-D sketch

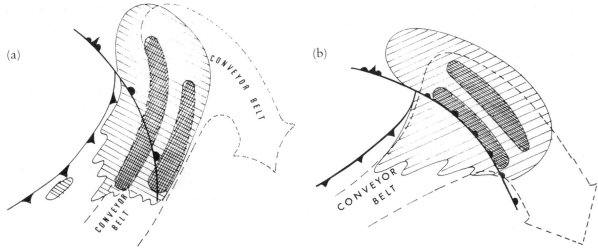

(a)

(b)

38. (a) and (b) Rain areas associated with warm conveyor belts

39. Conveyor belt ascending
cold front:

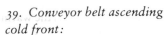

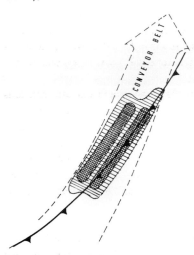

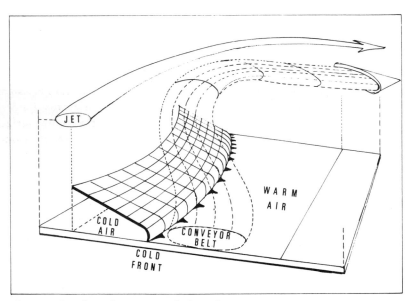

(a) plan view with rain areas   (b) 3-D sketch with conveyor belt and jet stream

Warm and cold conveyor belts

Cold conveyor belts are found transporting long tongues of cold polar air from high latitudes towards the developing depression. The cold conveyor belt passes underneath the warm belt. The cold current may then curve round the centre of the low and tuck in behind the cold front, or turn back and ascend beneath the warm current.

The second situation is illustrated in a rather complicated diagram—fig. 40.

The warm conveyor belt approaches from the south (outlined by full lines). The cold conveyor belt is shown by pecked lines starting from the eastern side of the low. The fronts are shown by the usual symbols for warm and cold and there is a short occlusion from the tip of the warm sector. A scalloped line with edge shading marks the boundary of the main frontal cloud mass.

Along the centre line of each conveyor belt are a series of numbers in circles. These indicate the top of the conveyor belts in thousands of feet. The warm conveyor belt starts with a top at 3,000 feet. This rises to 6,000 feet ahead of the cold front and to 10,000 feet in the warm sector just before reaching the warm front. The ascent is much steeper above the front; encircled values show tops of 14,000 rising to 30,000 feet where the current curves eastward ahead of the front.

The cold conveyor belt starts with a top of 6,000 feet and runs beneath the warm front until it has passed the tip of the occlusion (where the top is shown as 10,000 feet). Then the cold current makes a sharp right-hand bend as it ascends through 14,000 and 16,000 up to 18,000 feet. This extends the mass of rising air and shows that ascent is not always confined to the region above the warm front.

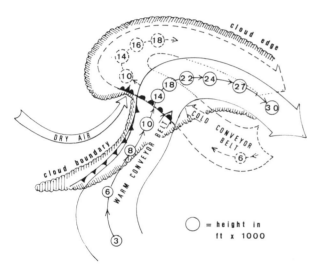

40. *Complex example of warm and cold conveyor belts showing how they overlap and rise over the fronts*

Sequence of weather as a frontal system passes

The sequence described below is typical of a well-developed frontal system.

Cloud

(a) **Cirrus (Ci).** The first signs of an approaching warm front are streaks of high cirrus, usually at levels between 30,000 and 40,000 feet. The streaks are sometimes called 'mares' tails'. The first cirrus may be more than 500 miles ahead of the front. If you are near an airway you may notice that the 'contrails'

(condensation trails) left by high-flying jets start to become thicker and persist much longer. This is due partly to an increase of moisture at high levels.

As the front approaches the streaks of cirrus merge to form an almost uniform sheet of **cirrostratus (Cs)**. The ice crystals in this cloud refract the light and produce a halo of 22° radius round the sun or moon.

(b) **Altostratus (As).** The Cs often merges imperceptibly with an As sheet as the cloud lowers towards the front. The As clouds consist of droplets of supercooled water with temperatures far below freezing. These clouds usually look grey; they do not give a halo round the sun or moon.

The first precipitation, small crystals of snow, fall from the As sheet. This snow usually evaporates high up because the air below the frontal surface is usually very dry at high level. However, the snow may be observed by radar a hundred miles or more before the main rain belt.

(c) **Stratus (St)** and **nimbostratus (Ns)**. As the warm front comes closer the medium level As cloud lowers. When rain falls out of it the cloud is called 'nimbostratus'. (Nimbus is the Latin name for rain cloud.) The rain moistens the air below the front and layers of stratus cloud often develop at low levels well below the frontal surface.

When the front is very close the low stratus thickens to form a continuous layer of cloud which covers high ground and may fall to the surface. In the warm sector (after the passage of the warm front) the low stratus tends to persist but the higher layers of cloud gradually break up and disperse. In summer the low stratus often breaks up well inland, especially in areas sheltered by a range of hills.

(d) The approach of an active cold front is often preceded by the appearance of a medium-level layer of cloud called **Altocumulus (Ac)**. This is distinguishable from the uniform sheet of As ahead of a warm front by its cellular structure. The Ac layer tends to have rolls or ripples in it and may have cumulus-like tops.

(e) At an active cold front there is likely to be a line of towering **cumulus (Cu)** or **cumulonimbus (Cb)**. From high levels the Cb line can be seen from a long distance. On closer approach the line may be seen to be broken into short segments allowing a flight through narrow gaps between the Cb.

To an observer on the ground there is often little advance warning because the warm sector low stratus may obscure the advancing line of Cb until it is very close. The Cb tops may never be seen from the ground but the line of the front is usually marked by a much darker line of cloud from which heavy rain falls.

The belt of cold frontal cloud is usually much narrower than the warm-front cloud. After the Cb belt has passed the cloud base lifts and the low stratus disperses as colder air arrives. There may be further outbreaks of rain before the frontal cloud lifts and breaks.

(f) **Post frontal Cu.** When the cold front has passed the air becomes colder and drier and cumulus cloud develops. There is often a clear gap before these Cu appear. This is because the front is often followed by a zone of descending ('subsiding') air which prevents any Cu from developing. When the zone of subsidence has passed the Cu start to develop and may grow into Cb with showers.

Weather

As the warm front approaches snow starts to fall from cloud above the freezing level. Below the freezing level the snow turns to rain. There is a shallow band where snow and rain exist together producing a strong echo on radar. This appears as an almost horizontal band on vertical scanning displays and is called the 'bright band' because it shows up so clearly.

The rain is usually heaviest near and just ahead of the warm front, but bands of moderate to heavy rain also develop well in advance of the front. After the front has passed the rain usually turns to drizzle, indicating that the much deeper rain-bearing clouds have passed.

At an active cold front there is usually a belt of heavy rain, sometimes with thunderstorms and squalls. This is only likely if the conveyor belt is rising up the frontal surface. A kata cold front seldom gives more than light rain or drizzle.

Visibility

Visibility normally decreases as the frontal rain arrives and may fall to a mile or less in thick drizzle. Hill fog with visibilities of less than 100 yards occurs when the low stratus settles down on high ground. Sea fog may produce equally bad visibility when the dew points in the warm air are higher than the temperature of the sea.

After the cold frontal rain has passed the visibility often becomes extremely good, especially when the

air behind has come from polar regions. Gales, which make the sea very rough, often carry a layer of spray and salt particles for long distances and this causes a shallow layer of reduced visibility.

Wind

As the warm front approaches, the surface wind generally backs and usually increases in strength. For example, in the United Kingdom and north-west Europe the wind direction may change from NW to W and continue backing to become southerly as the front approaches. When the warm front passes the wind nearly always veers, from S to become SW for example. The direction then tends to be fairly constant until the cold front approaches. Cold fronts, especially active ones, often have a belt of much stronger winds just ahead of them. As the front arrives the wind speed often reaches its peak to produce a line squall. Much of the destructive effects of a gale are concentrated along this squall line.

Passage of the cold front is marked by a veer of wind, from SW to WNW for example, and after that the wind speed usually decreases. The actual wind direction varies with each system but the winds invariably veer after the passage of a front.

Temperature and dew point

The air temperature near the surface is not always a reliable guide. At 2,000 feet or above there is nearly always an increase of temperature when the warm front passes and a decrease after the cold front. At the surface these changes may be masked.

If it has been sunny overland before the warm frontal cloud spreads over, the temperature near the ground may actually fall as the cloud and rain arrives. If it becomes sunny after the passage of the cold front the surface temperature may become warmer in the 'cold' air than it had been in the 'warm sector'.

The dew point is a much better guide. It depends on the amount of water vapour held in the air. High dew points mean there is a lot of water vapour. The dew point always rises when the warm front passes through and stays high throughout the warm sector. It may reach its highest value just before the cold front arrives. Beyond the cold front the dew point decreases and the change can be very marked.

Summary

The approaching warm front is marked by thickening cloud, starting with Ci at high levels and lowering to Ns in rain near the front. The wind backs and freshens, the dew point rises and the cloud base and visibility falls. At the warm front rain turns to drizzle, the cloud base is lowest, the dew point stops rising, and the wind veers.

In the warm sector the dew point is highest, drizzle and very low cloud with hill fog are common, especially near windward coasts and hills.

When an active cold front arrives there is a band of heavy rain, sometimes thundery. The wind increases to a maximum and veers sharply as the front goes through. Then the dew point falls, the cloud base rises and visibility improves.

Instability above a warm front

Warm frontal clouds are often described as being stratiform: cirrostratus, altostratus and nimbostratus. This is not universally true. As the warm conveyor belt rises above the frontal surface the air cools and in some circumstances may become unstable. One kind of instability (described in more detail in the next chapter) is called 'potential instability'. This occurs when the air starts off with a high moisture content before lifting begins. If a drier current spreads over at a higher level the result of prolonged lifting results in a saturated and unstable air mass. Then instead of stratiform clouds cumulonimbus may develop. The Cb are embedded in the layer clouds so they can only be detected by the radar returns or seen from much higher up.

Figure 41(a) shows a cross-section illustrating the ascent of a warm conveyor belt (shown shaded) over a warm front. At higher levels a current of much drier air is flowing in. This combination is potentially unstable; the instability is realised when the conveyor belt has risen far enough. The result is illustrated in fig. 41(b). Cb have developed and are embedded in the stratiform layers with only their tops protruding. These Cb produce areas of heavy rain near and ahead of the warm front; in summer violent thunderstorms may occur.

The warning signs are:

(a) Very high dew points reported by stations under a warm conveyor belt where the winds are fairly strong from a southerly direction. Dew points of 18–20°C suggest a high risk of thunderstorms.

(b) A veer of winds with height so that dry SW to W winds spread over the very moist conveyor belt.

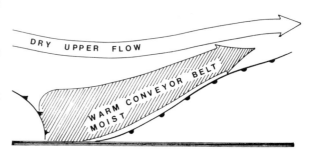

41. *(a) Conveyor belt ascending warm frontal surface with dry upper flow above it leading to potentially unstable air*

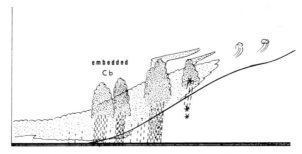

(b) Cloud structure showing how Cb develop through warm frontal layer cloud when potential instability is realised

A split cold front

Figure 42 shows a situation which gives rather similar weather. In this example the upper current has descended above the original cold front, making it a katafront, and gone on to push cold air aloft over the warm conveyor belt ascending the frontal surface. This structure has been called a 'split cold front' because the cold air has split into two sections. The upper part has gone ahead and begun to override the warm conveyor belt; the lower part has been left behind. The warm front cloud thus has the upper part of a cold front on top of it. There is then a region of heavy, perhaps thundery rain ahead of the warm front with a band of Cb at medium and high levels embedded in and extending above the Ns and As.

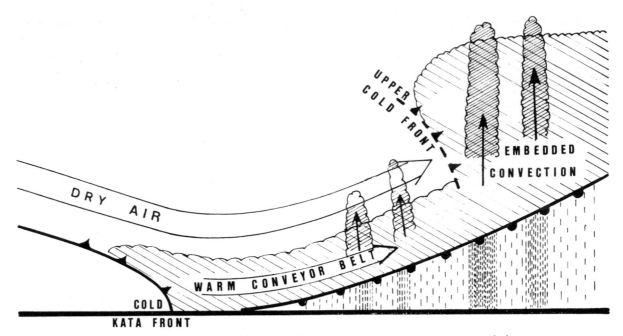

42. *Cross-section of a 'split cold front' showing cold air over running warm conveyor belt*

This combination makes flight through the warm front particularly rough. There may be severe turbulence, heavy icing and thunderstorms. Meteorologists often find such systems difficult to analyse correctly because the system seems to be a mixture of warm, cold and occluded fronts. A good coverage of upper air soundings combined with infra-red satellite pictures is needed to find out what has happened.

The satellites show that the deep cold frontal cloud has been transferred far ahead of the surface front. The original cold front, now a katafront, still has high dew point air ahead of it causing drizzle, very low cloud and hill fog.

It can be very puzzling to a ground observer. There is a period of heavy, even thundery, rain but the low cloud does not lift and clear until the lower part of the split cold front goes through. Pilots flying higher up can see a line of Cb which looks like the cold front but the low cloud does not break as expected. There may be little to mark the lower part of the cold front when you fly over it, just some breaks in the Sc further west.

Other cold front variations

The cloud on some cold fronts consists of a series of short segments rather than a continuous line. This can be seen when flying high above the front or by looking at detailed radar pictures. Figure 43(a) shows the front as it appears on a Met. chart. The observations are too far apart to reveal any irregularities and so the analyst draws it as a continuous line where the isobars show a sharp kink. Figure 43(b) shows the radar picture. The cloud now appears as a series of short lengths which are not exactly parallel to the general line of the front. Each section is at a small angle to the line of the front.

Figure 43(c) is an enlarged analysis. It suggests that the front is really a set of fairly straight lines joined by short bendy sections. The main build-up of cloud occurs along the straights with gaps at each bend. The

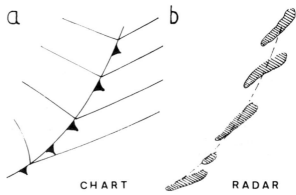

CHART   RADAR

43. *(a) Cold front as shown on a Met. chart*
(b) Cold front as observed by radar

wide arrow to the left shows the cold air moving at 15–20 knots from the NW. A single, thin arrow shows that the front itself is moving from the WSW. The much wider arrow marks a feature known as the 'low-level jet'.

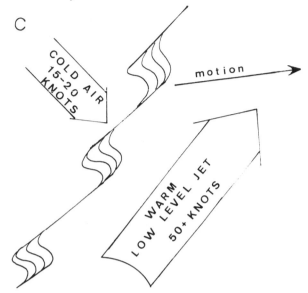

43. (c) Enlarged plan of cold front

There are several kinds of low-level jet; this one is a belt of strong winds that lies just ahead of the cold front. The speed in a low-level jet is usually less than half the speed of a high-level jet but is strong in relation to the surrounding winds. This low-level jet lies in the warm conveyor belt bringing moist tropical air into the warm sector of a depression.

Figure 44 is a simplified 3-D view of fig. 43(c). The leading edge of the cold front has the indentations shown in the previous plan view. The front itself is almost vertical near the surface and here the warm air is forced up very rapidly to produce lines of Cb. The frontal zone is extremely narrow in this region, unlike some warm fronts which have a very wide transition zone between cold and warm air. A strong downdraft often occurs in or close to the heavy rain at the front.

The updraft at the nose of the front and the downdraft close behind can make the region very turbulent. Anyone trying to fly through below cloud is liable to have a very rough ride.

Figure 45 shows a more elaborate 3-D sketch of a similar situation with cloud and rain added. The airflow is quite complicated. There is a rising current which curves over the top of the low-level jet and a

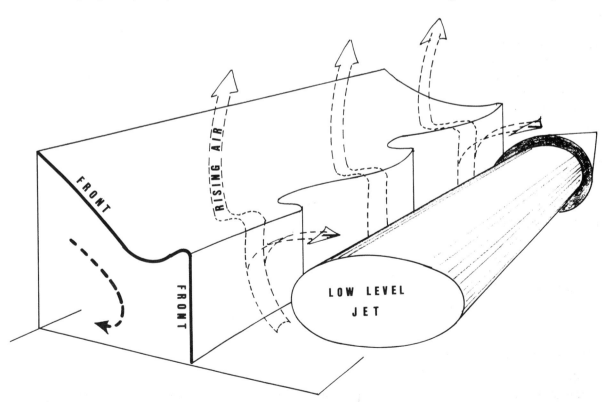

44. Schematic 3-D structure of the cold front showing low-level jet ahead of it

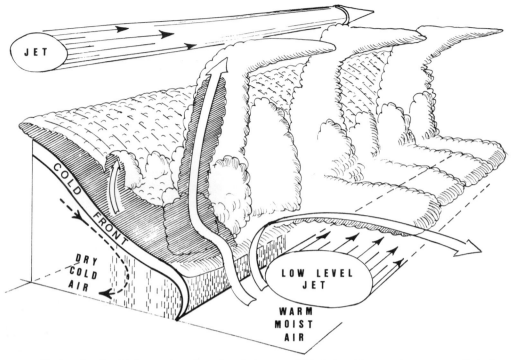

45. 3-D sketch of cold front showing cloud, low- and high-level jets and patterns of air flow

more powerful up current forming the Cb. Beneath the frontal surface cold, dry air from high levels descends towards the surface. Near or beyond the western edge of the frontal cloud lies a high-level jet blowing at a small angle to the line of the front.

The changeability of fronts

The diagram and descriptions above have illustrated several varieties of front but they are by no means the full story. Every year new features are discovered by people who probe these systems with doppler radar, balloons, aircraft and satellite. Here are some principles which generally prove true.

1. The nearer you are to the centre of the low the worse will be the frontal weather. As you fly well away from the depression the cloud tops usually become lower, or at least more layered, but the cloud base may still be very low.

2. If the Met. charts show a sharp kink in the isobars the front will probably be more active there.

3. The worst conditions on a warm front are often near windward coasts. In summer if the front travels well inland the cloud base generally rises during the day. If the front has to pass over mountains most of the low cloud will be held up on the windward side.

4. Cold frontal weather can be very severe when there is very warm moist air just ahead of it. The extra warmth and moisture increase the energy released by the frontal Cb producing severe thunderstorms, squalls and torrential rain, turbulence and a very low cloud base.

5. If the warm sector air has a very high dew point a line of thunderstorms may break out far ahead of a cold front. This is almost invariably a summer event which takes place over very hot land during the afternoon and evening. When a pre-frontal thunderstorm line forms, one often finds the cold front itself arrives later as quite a weak feature.

6. When a winter cold front moves in from the sea over a much colder continent the front behaves more like a weak warm front at low levels.

7. Beware of a front which stands still. The cloud tops may not rise far but the base often sinks very low in drizzle.

4

Stability in the atmosphere

Clouds appear in a variety of forms which are changing all the time. Time-lapse films show how cumulus clouds surge upwards and then collapse; streaks of cirrus embedded in a jet stream rush across the sky; stratus undulates gently up and down like the tide in a river estuary. The wind drives the clouds along but the stability of the air controls the shape of the clouds. Stability in the atmosphere is usually described in terms of 'lapse rates'.

Lapse rates

The lapse rate defines the way in which temperature varies with altitude. It is usually quoted under two main headings:

(a) The environmental lapse rate (ELR) is the actual variation of temperature with height at a certain time and place. The measurement is usually carried out by releasing a balloon which lifts a small instrument package called a 'radiosonde'. The radiosonde measures pressure, temperature and humidity as it rises through the atmosphere and transmits the readings to an observer on the ground or at sea. Research aircraft reverse this process and use a 'drop sonde' which descends by parachute.

(b) The adiabatic lapse rates. The word 'adiabatic' means that no outside heat reaches the volume of air. Air is a very poor conductor of heat and (unless there is turbulence to stir up the whole mass) it takes a very long time for any outside heat to affect the central regions of a large volume of air. The temperature changes which do take place are almost entirely due to internal variations of pressure.

If a parcel of air is lifted it rises into a region of lower pressure. The rising parcel adjusts to this lowered pressure by expanding until the pressures inside and outside are equal. The process of expansion uses up some internal energy; this energy is taken from the heat in the air so the temperature falls. The rate at which temperature changes is called the 'adiabatic lapse rate'.

The dry adiabatic lapse rate (DALR)

This is the rate at which the temperature of a parcel of dry air is reduced when it is moved up, or increased when it descends. The value is 9.8°C per kilometre or 3°C per thousand feet. This lapse rate remains constant until the air is saturated with moisture.

The saturated adiabatic lapse rate (SALR)

The air can only hold a limited amount of water vapour, depending chiefly on the air temperature. Warm air can hold more water vapour than cold air. When the temperature is reduced there comes a point when the air is saturated and unable to hold any more water vapour. Further cooling results in condensation of the vapour into tiny droplets of water which form cloud or fog in the atmosphere or dew upon a cold surface. This temperature is called the 'dew point'.

Latent heat

This is released by the process of condensation. The release of heat warms up the air parcel. The further the air is raised, the greater is the amount of heat released by condensation. This reduces the rate of cooling when saturated air is lifted; the SALR is therefore less than the DALR. In the lower levels of the atmosphere the SALR varies from about 1.2°C per thousand feet when the temperature is 26°C, to 2.2°C per thousand feet in cold air with a temperature of −10°C. With prolonged lifting less and less water vapour remains to be condensed. As a result, the supply of latent heat dwindles as the temperature falls and eventually there is not enough to have any measurable effect on the lapse rate. At high levels where the air temperatures fall below −40°C the SALR and the DALR are almost identical.

Graphical representation of lapse rates

Figure 46 shows a graph with temperature shown along the base line and height marked up the side. The straight parallel lines are dry adiabatics with a slope (lapse rate) of 3°C per thousand feet. The pecked lines are saturated adiabats. These are slightly curved because the SALR is not constant. Diagrams like this can be used to find out whether the air is stable or not.

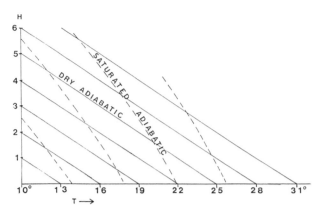

46. *Temperature height diagrams*

A definition of stability

Air is stable if it returns to its original level after an initial vertical displacement. A parcel of air may be called 'buoyant' if it tends to rise through its surroundings. Buoyancy depends on density; if the air is less dense than its surroundings it will tend to rise, if it is more dense it will sink. Now density depends on the mass of air within a given volume. If this air is heated it will expand to fill a larger volume of space and in doing so becomes less dense.

When parcels of air at the same pressure are compared the difference in densities depends on the difference in temperature. The warmer, less dense, parcel will be buoyant and tend to rise through the colder surroundings. The rising air is called a 'thermal'.

Thus, a parcel of air is unstable and will tend to rise so long as it remains warmer than its surroundings. To find out how far the parcel can rise we need to compare its temperature at each stage with the environment. Table 4 below shows such a comparison.

Table 4. A comparison of the temperature of a parcel of rising air with the environment temperature

| Height | Temperatures (Deg C) | | |
	Environment air	Parcel air	Difference
Surface	25	29	+4
1,000 feet	23	26	+3
2,000 feet	21	23	+2
3,000 feet	19	20	+1
4,000 feet	17	17	0
5,000 feet	15	14	−1

This table shows an environmental lapse rate of 2 degrees per thousand feet compared to the dry adiabatic which is 3 degrees. The surface air has been warmed by contact with a hot ground and starts off 4 degrees warmer than its environment so (being less dense) it will rise. As it rises the difference in temperature is reduced by 1 degree per thousand feet and at 4,000 feet it is the same as the environment. If it had enough momentum to continue rising it would be 1 degree colder at 5,000 feet. Then, being colder and denser than its environment, it would sink back to 4,000 feet.

If there was no other factor involved one could predict that a 'dry thermal' starting off with an excess of 4 degrees at the surface would rise to 4,000 feet in this particular environment.

Adding moisture

In the next example the air, instead of being completely dry, contains a proportion of water vapour such that dew would form on the ground if the air was cooled below 19°C (i.e. the dew point is 19). The temperature height graph can be adapted to show how the dew point line slopes with altitude.

This is illustrated in fig. 47. A line has been drawn upwards from the surface dew point of 19; it meets the dry adiabat at approximately 4,000 feet. At that point the temperature of the rising parcel is the same as the dew point; the air is then saturated and further cooling will lead to condensation of cloud droplets. This point is known as the 'condensation level'; it is also (very nearly) the base of any cumulus cloud. Above the condensation level the air will cool off at the saturated lapse rate.

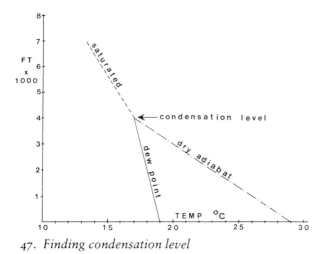

47. *Finding condensation level*

A rule for finding cloud base

Figure 47 illustrates a useful rule for estimating the cloud base during the morning and early afternoon when sunshine is warming the ground and the air in contact with it.

Multiply the difference between the air temperature and the dew point by 400 to find the cumulus base in feet. In fig. 47 the values are 29 and 19 giving a difference of 10 degrees. Multiplied by 400 this gives a cloud base of 4,000 feet. The rule does not work later in the day when the temperature starts to fall. It has a limited value when there is a fresh breeze creating turbulence to stir up the air. Then it is usually true to say that the cloud base will not be below this height, though it may be well above it.

For example, if there is a wind of 15–20 knots at the surface and the temperature difference is 5 degrees, the cloud base is unlikely to be lower than 2,000 feet (5 × 400) and may be much higher. The air temperatures and dew points for certain major airfields are given in 'Volmet' broadcasts. These are detailed in the appendix.

How cloud development affects stability

Table 5 shows a modification of Table 4. The environmental lapse rate has been extended up to 8,000 feet and a temperature inversion introduced at 6,500 feet. The surface dew point has been raised 1 degree bringing the condensation level down to 3,600 feet.

Table 5. A comparison of the temperature of a rising parcel of air with the environment temperature when cloud forms

	Temperatures (Deg C)			
	Environ-ment	Dry parcel	Wet parcel	Excess temp
Height	air	air	air	
Surface	25	29	—	+4
1,000 feet	23	26	—	+3
2,000 feet	21	23	—	+2
3,000 feet	19	20	—	+1
3,600 feet	17.8	18.2	18.2	+0.4
(cloud forms at this level)				
4,000 feet	17	—	17.7	+0.7
5,000 feet	15	—	16.5	+1.5
6,000 feet	13	—	15.3	+2.3
6,500 feet	12	—	14.7	+2.7
(inversion above this level)				
7,000 feet	13	—	14.1	+1.1
8,000 feet	15	—	12.9	−2.1

This shows that the change from dry to saturated lapse rates at 3,600 feet (when the cloud formed) made the air unstable up to about 7,300 feet instead of 4,000 feet. It was only the presence of an inversion above 6,500 feet which stopped the cloudy air rising even higher (see fig. 48).

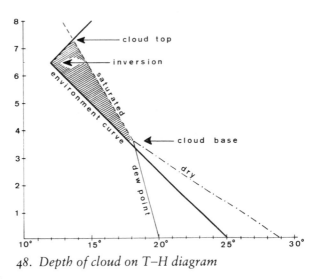

48. *Depth of cloud on T–H diagram*

The importance of inversions

A temperature inversion occurs when the lapse rate is the opposite (inverse) of normal, that is when the

temperature increases with height instead of decreasing. This makes the layer very stable. Inversions act as a 'lid' restricting the height to which warm air can rise. The base of an inversion is often found near or just below the tops of cloud or fog. The rising columns of cumulus cloud often spread out to form a layer when they encounter an inversion. During periods of poor visibility the top of a haze layer usually coincides with the inversion level.

Defining stability in relation to lapse rates

The examples given suggest how stability can be defined in terms of lapse rates:

Absolute instability

This is when the environmental lapse rate (ELR) is greater than the dry adiabatic lapse rate. In other words if the temperature aloft decreases at more than 3°C per thousand feet. This is called a 'super-adiabatic' lapse rate. It normally only occurs in a shallow layer

The top of cumulus spreading out where it meets an inversion at 10,000 feet. The inversion was just developing when a ridge moved in after a showery spell

near the ground which has been made very warm by strong sunshine.

Neutral stability

This is when the environmental and the dry adiabatic lapse rates are the same (both 3°C per thousand feet). This is the rate at which a 'dry thermal' cools during its ascent from ground level to cloud base. It is called neutral because air moving up or down keeps its initial momentum and does not accelerate or slow down.

Conditional instability

This is when the environmental lapse rate is less than the dry but more than the saturated lapse rates. The ELR in fig. 48 was conditionally unstable. It only became unstable on condition that the air was saturated.

Absolute stability

This is when the environmental lapse rate is less than the saturated lapse rate. For example, if the temperature only decreased by 0.5 degrees per thousand feet the air would be absolutely stable because its lapse rate is less than the saturated rate.

Potential instability

This is a condition when the air is moist at low levels but dry higher up. It is not always obvious from a brief glance at an aerological diagram. The criterion is that the wet bulb temperature must decrease at more than the saturated adiabatic lapse rate. The potential is only realised if the whole air mass is lifted bodily, for example if a large mass of air was forced to rise up a long slope such as a front until saturation occurred. The process is described at the end of this chapter and the importance of potential instability is explained in the sections on fronts and cumulonimbus.

Factors which change the stability

In the examples shown in figs. 48 and 49 the original cause of instability was the heating of the surface air by contact with a warm ground, so that it started off with a temperature difference of 4°C over its environment. This instability occurs over land during sunny days and also over the ocean when cold air passes over a warm sea. The heated surface air is carried up by convective currents called 'thermals'. The structure of thermals is described in more detail in Chapter Five.

The height to which these thermals can rise depends on the environmental lapse rate. This seldom remains constant for long. Large masses of air are made to ascend or descend where the wind flow converges or diverges.

Subsidence and its effect on stability

When a large mass of air descends slowly it is said to be 'subsiding'. In the region of a developing anticyclone the air aloft may descend some 3,000 feet in a day. Descending air is compressed and warms up as it subsides. The warming first evaporates any cloud droplets and when the air is no longer saturated the temperature rises 3°C per thousand feet (the DALR). Thus, a descent of 3,000 feet might warm the air by 9°C in a day. The subsiding air mass cannot sink through the ground; near the surface the air spreads

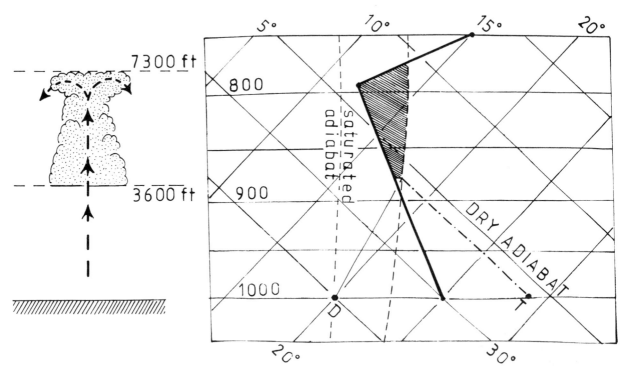

49. *Illustrating convection and cloud formation on a tephigram*

out sideways in the region of divergence. As a result the lower air hardly descends at all and its temperature rises very little; nearly all the warming occurs higher up. This commonly causes a temperature inversion above the ground.

Large scale ascent

If a large volume of air is forced to rise it starts to cool at 3°C per thousand feet until saturated; after that the cooling follows the appropriate saturated adiabat. Prolonged ascent tends to reduce the stability of the air, especially if the air was moist at low levels but much drier aloft. Such a condition may be potentially unstable. Many days which ended up with widespread thunderstorms started out with air which appeared stable to surface heating, but was potentially very unstable when lifted a few thousand feet.

Aerological diagrams

An aerological diagram is a graph on which observations of pressure, temperature and humidity aloft may be plotted. At first sight most aerological diagrams look like a complicated criss-cross of lines. Many people can get on very well without bothering with them. They should skip the rest of this chapter. It is only written for readers who want to dig deeper into the subject.

Most aerological diagrams show dry and saturated adiabatics as well as lines for pressure and temperature. This enables one to judge the stability of the air by comparing the actual values of temperature and pressure with the nearest adiabat. Many diagrams are designed to emphasise the difference between lapse rates and to represent the energy involved in different processes by areas on the diagram.

The tephigram

This is the aerological diagram used by the British Met. Office. At first sight it looks dauntingly complicated but so do many wiring diagrams and yet people manage to find their way round them.

The name comes from 'T' for temperature and 'phi' the Greek letter used for entropy in certain Met. equations. The word entropy has a range of definitions in the dictionary; to the Meteorologist it is a function of the pressure, volume and temperature of the atmosphere. On the tephigram a line of constant entropy is the same as a dry adiabatic. Moving along a dry adiabat the pressure, volume and temperature of a parcel of air alter with height but the entropy remains constant. Motion up or down an adiabat is called 'isentropic' (constant entropy).

The axes of a tephigram

The axes are Temperature and Entropy; the isotherms and dry adiabats are at right angles. As a result the pressure lines (isobars) run diagonally across the diagram and are slightly curved.

Figure 50(a) shows the basic structure of a tephigram. The horizontal lines are the dry adiabatics (and also lines of equal entropy). The vertical lines marked 0, 10, 20 show the temperature in °C. The isobars (lines of equal pressure) run diagonally. They are labelled in millibars from 1,050 up to 800 mbar (a height range of approximately 7,400 feet in this example).

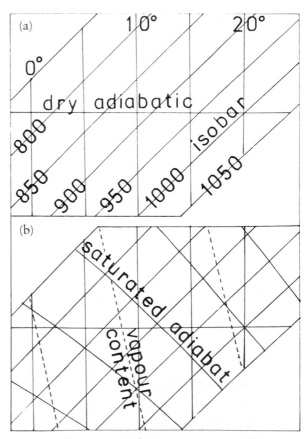

50. *Basic lines on a tephigram*

Figure 50(b) shows two additional sets of lines. The saturated adiabatic lapse rates (SALR) curve upwards from right to left. The pecked lines sloping up at a much steeper angle represent the vapour content; they also show how the dew point varies with height when a parcel of air is lifted.

Figure 49 shows how the example in fig. 48 looks when the values are transferred from a simple pressure height diagram to a tephigram. The heavy black line is the environment temperature. To avoid clutter only two saturated adiabats are shown and one vapour content line rising from point 'D', the surface dew point. The surface temperature of 29°C is marked on the 1,000 mbar line at point 'T'.

The condensation level is where the dry adiabat from T and the vapour content line from D meet. At that point the rising air becomes saturated and from then on the air cools at the saturated adiabatic rate.

The region between this saturated adiabat and the environment curve has been shaded in. The area of shading represents the extra heat energy provided by condensation. When this shaded area is wide the cloudy air is particularly buoyant and may contain strong upcurrents. A very narrow shaded zone suggests only slight buoyancy and weak upcurrents. The higher this shaded zone extends, the taller are the clouds.

On the left of this skeleton tephigram is an illustration of air rising from the surface, condensing to form a cumulus cloud based at 3,600 feet, and curling over sideways when it penetrates the inversion. This limits the cloud top to about 7,300 feet.

Figure 51 shows an example of warming by subsidence. The heavy black line is the original temperature curve. Arrows show how the air warms as it descends. The shaded area represents the temperature change. The final result is a temperature inversion from the surface up to the 900 mbar level. Inversions are very stable layers and inhibit the development of thermals. The air over an anticyclone often shows this strong stability at low levels. It suggests hot sunny days in summer but cold foggy nights in winter.

The release of potential instability

Figure 52 shows an example of potential instability. The original temperature curve is marked with a thick black line. Arrows show how this mass of air

cools during large-scale lifting. At the bottom, where the air is moist, it soon becomes saturated and further ascent produces cooling at the SALR. Higher up, where it was initially drier, it cools off much more before saturation occurs. What happens during ascent is that the air low down cools much less than the air high up. The line marked 'after lifting' shows the extent of the change. At the bottom the cloudy air

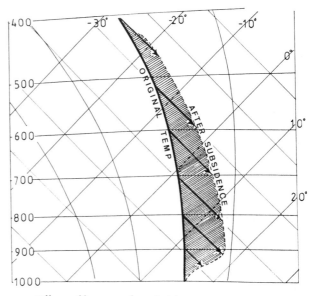

51. Effect of large scale subsidence

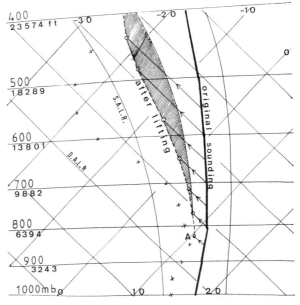

52. Large scale lifting and potential instability

tends to rise along a saturated adiabat. The shaded region shows the energy available for ascent.

When the environmental temperature is so modified by ascent cumulonimbus clouds may develop to great heights. These Cb clouds are often embedded in other cloud layers. This makes conditions treacherous for aircraft without cloud-warning radar. One may be flying through relatively smooth layer cloud and suddenly encounter severe turbulence, heavy rain, perhaps hail and thunderstorms.

Figure 53 shows the lowest section of a tephigram. The fundamental lines, isotherm, dry adiabat, saturated adiabat and dew point line are labelled. The slightly curved isobars are labelled every 50 mbar from 1,050 mbar up to 750 mbar. A full-sized tephigram would be extended up to about 50 mbar. The heights (in feet) indicated by an altimeter set to 1,013.2 mbar are given at the left-hand side at intervals of 50 mbar. Thus, 1,000 mbar is equivalent to 364 feet and 750 mbar to 8,091 feet.

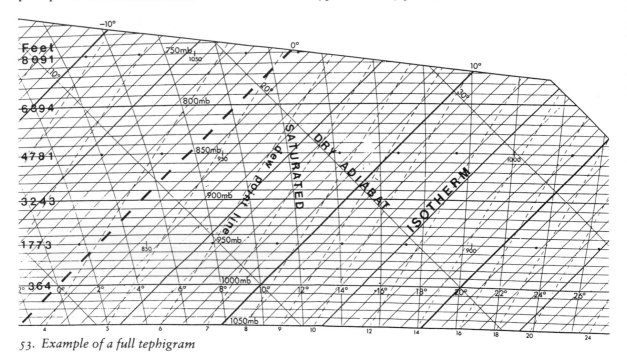

53. *Example of a full tephigram*

5

Convection and cumulus clouds

Surface heating and the rise of temperature

When sunshine heats the ground the air in contact with it is also warmed. The warm air becomes less dense than the environment and starts to rise. The rising air currents carry the heat upwards; this redistribution of heat is called convection. The depth of convection depends on the original lapse rate of temperature and the amount of heat supplied at ground level.

Figure 54 illustrates how the temperature changes as convection proceeds. The diagram consists of three pairs of figures. The upper parts show simplified tephigrams while the lower graphs show how the surface temperature and the top of the convective layer rise during the day. On these graphs time is shown along the base line. Temperature is marked on the left-hand side and the height to which thermals rise on the right-hand side.

Figure 54(a) shows an isothermal atmosphere. The thicker black line along the 12°C isotherm represents the conditions before sunrise. As the surface temperature rises the lapse rate is changed from being isothermal to dry adiabatic. In this example, which is for a day in the middle of June in central England, the temperature reached a maximum near 24°C. At that temperature the dry adiabat extends to nearly 4,000 feet. The shaded area on the tephigram represents the energy supplied to make this change.

The lower curve on the graph below shows how the surface temperature rose hour by hour. The upper curve shows the top of thermals; the height which the DALR reached. On this date the sun rises at about 0345 hours and the air temperature usually starts to rise at about 0430 hours.

Figure 54(b) shows the effect of an inversion of temperature. The surface temperature starts at 12°C as before but increases with height to become 20°C some 4,500 feet higher up. The shaded sector of the tephigram has the same area as that in (a) but the thermals do not rise as high. In this example they stop

at about 3,000 feet. On the graph below the surface temperature can be seen to rise faster to reach its afternoon maximum of 26.6°C. The surface temperature rises higher because the inversion limits the depth of air through which the heat is distributed. Most people have noticed a similar effect when heating a building. A limited amount of heat warms a room with a low ceiling much better than a large church whose roof is far higher.

Figure 54(c) shows a more common situation. An

54. *Graphs of temperature rise for different lapse rates: (a)*

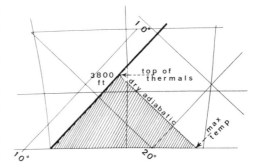

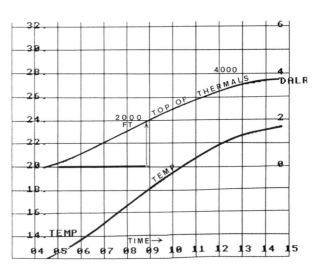

54. (b)

54. (c)

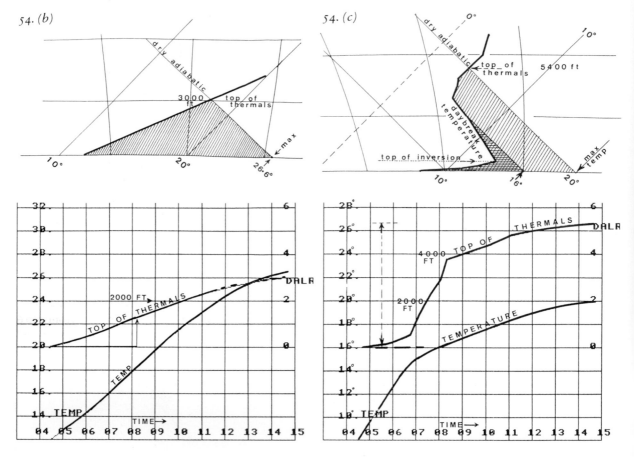

aircraft sounding made at daybreak found a very shallow inversion near the surface, then the temperature decreased with height. Just below 4,000 feet the air became more stable and from 5,000 feet upwards there was an isothermal layer. The single shading shows the total area of heating by mid-afternoon. The double shading shows the amount of heating achieved by 0800 hours.

The graph below shows the temperature curve rising very rapidly until the low-level inversion has been broken (by about 0700 hours). Once this inversion had gone the thermals could rise much higher. As a result the heat was distributed over a greater depth and so the surface temperature rose more slowly.

Notice that the upper curve, showing the height to which thermals can rise, jumps from a few hundred feet just before 0700 hours to nearly 4,000 feet soon after 0800 hours. After that the rising thermals encountered the stable layer aloft and the line marking the top of thermals began to level off. Graphs like this are very useful for predicting soaring conditions.

Figure 55(a) shows an enlarged (and much simplified) tephigram. The early morning upper air sounding shows an environmental lapse rate labelled E, F, G, H. E is the surface temperature at sunrise (12°C). F is the top of the low-level inversion, G is the base of a much higher inversion, and H is as high as that particular sounding went. There is a series of parallel horizontal lines representing dry adiabats. As the surface temperature rises, thermals carry the extra heat upwards along one of these dry adiabats. With a further rise in surface temperature the dry adiabats grow longer, indicating that thermals can rise higher. When the surface temperature reaches J (18°C) the thermal can rise to K (9°C). K represents the top of the dry thermal JK at that time. Since the temperature decreases 3°C per thousand feet on a dry adiabat the height of K must be 3,000 feet.

Figure 55(b) shows the same initial distribution of temperature but an extra item—the surface dew point of 10.7°C has been added. A dotted line marked 'dew point' is shown rising from the surface to meet the environment line at K. With a surface temperature of

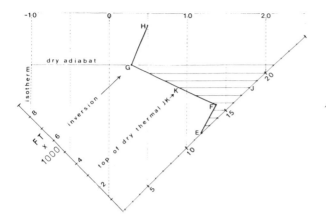

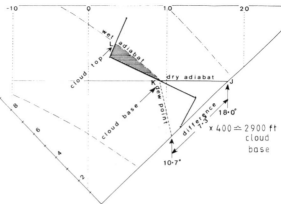

55. *(a) Showing how lapse rate changes as temperature rises*

(b) Determining temperature at which Cu will form

18 and a dew point of 10.7 the condensation level will be about 2,900 feet (400 × 7.3) which is just below point K. After condensation starts the thermal cools off at the saturated lapse rate (marked as a wet adiabat).

The shaded region represents the extra energy released by condensation. It would allow the top of a cumulus cloud to rise to point L; this is just above the inversion level.

The shape of thermals

Until a cumulus cloud develops there is nothing to show what a thermal looks like. However, it is possible to make a liquid model of a thermal using a glass-sided tank of water into which a cupful of liquid of different density is released. Early experiments used a salt solution (which is denser than pure water) to represent the thermal. The salt solution was made visible by adding a white precipitate. This is an upside-down thermal which sinks down instead of rising upwards. However, the motion in and around the liquid thermal is very like the motion one can observe in a growing cumulus cloud. Photographs of the sinking 'thermal' look (when inverted) so like real cumulus clouds that one is justified in assuming that the behaviour is essentially the same in water as it is in the atmosphere.

Water tanks are an excellent way of studying the behaviour of the atmosphere, provided that the scaling factors and differences of density are well matched.

Figure 56 shows some examples of thermals. (a) illustrates successive outlines numbered from 1 to 6 showing how the rising mass starts off as a column and ends up as a detached bubble cut off from the surface.

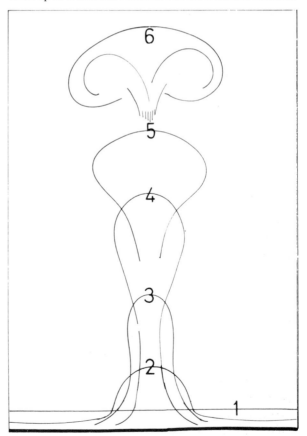

56. *(a) The shape of a rising thermal*

The shape of this thermal bubble is shown at (a). It resembles a vortex ring where the rising air ascends in the middle, turns outward at the top and starts to descend again round the sides. Glider pilots have confirmed that many real thermals often show similar characteristics. It is possible to climb very quickly in the central core but when you near the top of the bubble the rate of climb decreases. A second glider lower down can then catch up the first one.

The shaded part at the top of the thermal bubble in fig. 56(b) is a region where the warmer air in the thermal becomes mixed with the cooler air surrounding it. This mixing is an important feature of all thermals. Mixing entrains outside air into the thermal which expands and becomes diluted. This dilution makes the thermal less buoyant and eventually halts its ascent.

Figure 56(c) shows the ascent of a thermal over a period of time. The top of the thermal is outlined at regular intervals. These outlines show that protuberances which develop near the centre line tend to

56. (b)

56. (c)

spread out sideways as the top rises. Several pecked lines are added to show the outwards growth of various bulges.

Figure 57 illustrates how a simple thermal starts as a column (often called a 'plume') rising from the heated surface, develops a cap of cumulus cloud when it passes the condensation level, and turns into a bubble when the supply of warm air is exhausted. Flying through a thermal one usually encounters a slightly turbulent region at the edge where the air may actually be descending. Reaching the middle one flies through a column of 'lift'. Gliders normally start to circle at this point in order to gain height; if no circle is made the aircraft will soon be in sinking air outside the thermal. High-speed aircraft may only experience a jolt on passing through.

Figure 58 shows a similar development observed by instrumented motor gliders which made a number of traverses through thermals.

Figure 58(a) shows a large flat area (about the size of an airfield) over which there is a shallow layer of warm air resting on the heated ground. The lines numbered 1 to 5 show the potential temperature of the air.

Note. Potential temperature is the temperature which air would have if it was brought down a dry adiabat to the surface. Although air cools as it rises or warms as it descends the potential temperature remains constant so long as the air is dry. It is a useful way of identifying a parcel of air whose actual temperature changes when it moves up or down.

Figure 58(b) shows the layer of warm air starting to lift off the surface. A shallow layer of warm air may stick to the ground for some time until set in motion by some small disturbance. Once it begins to rise the bulge draws in warm air from all round.

Figure 58(c) is the stage just before the supply of warm air is exhausted. Lines of potential temperature identifying the thermal show that the central core is still warmer than its surroundings, but the temperature difference has been reduced. The thermal is still drawing on the supply of warm air from a large area of ground but the bottom of the thermal is becoming narrow. An extra feature is shown in this diagram; this is a stable layer aloft. In such a stable layer the potential temperature increases with height. The horizontal lines marked 5, 4, 3, 2 and 1 represent the stable layer.

At the stage shown in fig. 58(d) the thermal has bumped into the stable layer and some of the warmer

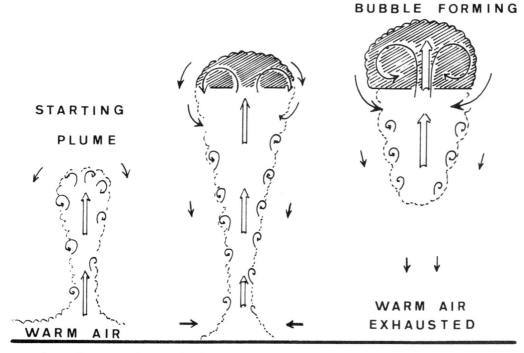

STARTING PLUME

BUBBLE FORMING

WARM AIR

WARM AIR EXHAUSTED

57. *Thermal rising to form a cloud*

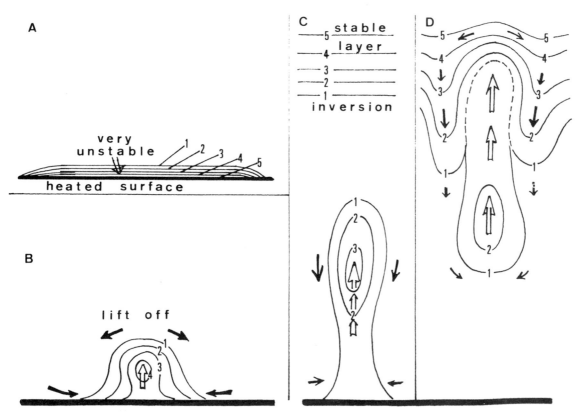

A

very unstable

heated surface

B

lift off

C   stable layer

inversion

D

58. *Temperatures in and around a rising thermal*

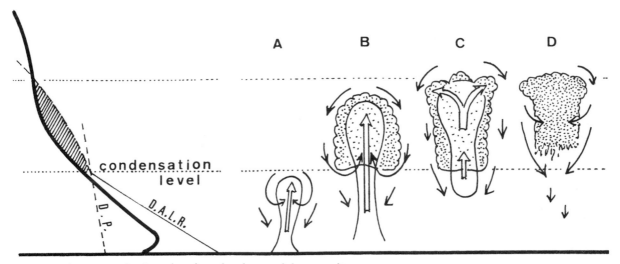

59. Development of a Cu related to the shape of the sounding

air has begun to sink down either side of the thermal. A rising thermal often has considerable momentum so it does not stop rising immediately. However, the air near the top of the thermal is now slightly cooler than its environment.

It is normal to find that the upper sections of thermals are no longer warmer than the surrounding air.

Figure 59 shows a combination of temperature sounding and corresponding cloud development. The temperature sounding is the heavy black line on the left. On it are drawn a DALR (dry adiabat from the surface temperature) and a dew point line (marked DP). Where these two meet is the condensation level and the shaded area represents the energy release when a cumulus cloud forms.

To the right, labelled from (a) to (d), is the life history of a thermal. At (a) it is a cloudless plume connected to the warm surface. At (b) it has risen above the condensation level and developed a cumulus cloud. The cloud provides extra energy so the thermal is strengthened. At (c) the supply of heat from the surface has been cut off but the cloud still continues to grow. Its top reaches more stable air and begins to spread out sideways. At (d) there is no longer any upcurrent in the cloud. The base decays and becomes ragged and sinking currents of air extend into the region previously occupied by the rising thermal. Eventually the entire cloud will disperse unless it moves over a new source of warm air.

Clouds which form over warm dry mountains often draw on a series of thermals rising up the heated sun-facing slopes. The cloud may then persist for a long time, the top can often be seen to pulsate as new thermals are drawn into it.

Dispersal of cumulus

There are days when cumulus develops in the morning but disperses later in the day even though the sun continues to heat the ground and produce thermals. Figure 60 shows one reason for the dispersal of cumulus by day. The temperature sounding on the left shows a big inversion limiting the cloud tops. When the surface temperature reaches A the dry adiabat A–A meets the dew point line (DP) and cloud forms in the lower shaded area. The illustration of Cu drawn for 'base A' shows what the clouds look like at this stage.

As the temperature rises a longer adiabat B–B produces a much higher cloud base (base B). Now the cloud is much shallower because the condensation level has almost reached the inversion. Finally, when the temperature rises to C the dry adiabat meets the environment curve before it reaches the dew point line. There is then no condensation and all that remains is a cloudless ('blue') thermal. The same process acts to disperse a layer of early morning stratus; it is called 'burning off'.

Figure 61 illustrates how cloud may be dispersed by a lowering of the inversion. Inversions are not fixed boundaries, they move up and down. Usually one finds the inversion rises on a warm sunny day but if an

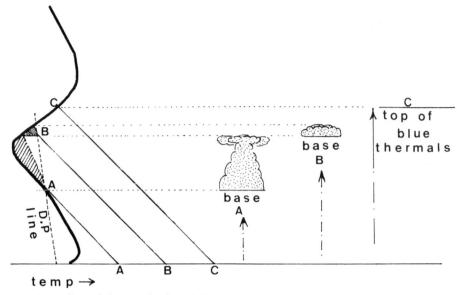

60. *Growth and dispersal of Cu below an inversion*

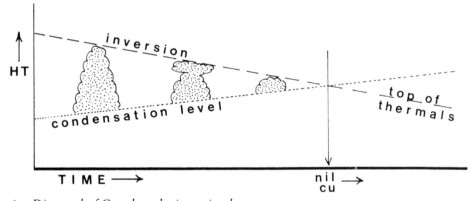

61. *Dispersal of Cu when the inversion lowers*

anticyclone is approaching the inversion may sink lower. The diagram shows two lines: the inversion which descends during the day and the condensation level which (nearly always) rises between morning and late afternoon. As the condensation level rises, the cloud base goes up. When the condensation level crosses the inversion level all the cumulus disperses. The point is marked as 'nil cu'.

Spreading out of cumulus

When high-pressure systems are centred over the sea it is common to find that cloud on the eastern side of the high tends to merge together to form an almost continuous sheet of stratocumulus. This cloud sheet is supplied with moisture by occasional cumulus clouds whose summits make small humps in the otherwise level surface.

The British Isles and much of western Europe may be covered by such a cloud sheet during the day.

Figure 62 illustrates the process. The sounding at the left shows a marked inversion acting as a 'lid' to limit the growth of any cumulus clouds. The usual two construction lines for dry adiabat and dew point show that the condensation level is well below the inversion. The air is very unstable beneath the inversion and although the morning may start with blue skies and excellent visibility the cloud forms unusually early and shoots up quickly. The air is normally very moist and sometimes cumulus tops are capped by a little lenticular patch of cloud called a 'pileus' (Latin for cap). Such pileus clouds form when

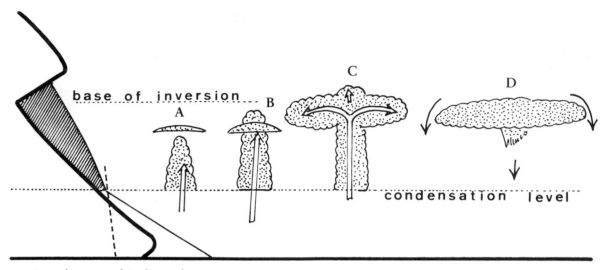

base of inversion

A   B   C   D

condensation   level

62. Spreading out of Cu beneath an inversion

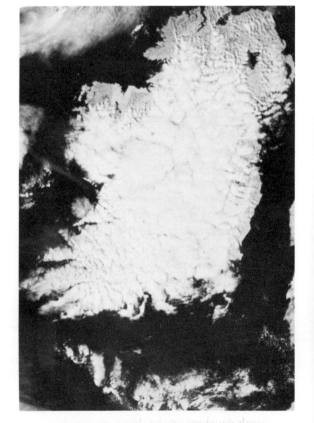

Stratocumulus formed by spreading out of cumulus. Most of this cloud is confined to the land because the sea was too cold to set off cumulus. The small cracks in the cloud layer over Ireland occur where the air sinks down at the edge of each cell
(Picture from Dundee University)

the rising cumulus pushes up some of the air above the thermal. This push is quite small; the cloud usually goes on to grow through its pileus. It shows that even a small amount of lifting can form cloud, indicating a very moist air mass at that level.

The sequence is shown from A to D. In A the cumulus is rather taller and thinner than usual and has formed its first pileus. In B it has risen through the pileus, still ascending fast. At C the cloud has bumped into the inversion and the rising air is spreading out sideways. Momentum carries a domed top a few hundred feet higher but nearly all the cloud is held down by the lid. At D the thermal has ended but the spread-out layer of stratocumulus persists.

The Sc layer cuts off the sun and once it has formed there are very few thermals. Lack of thermals may result in the slow dispersal of Sc; when the layer breaks up, the sun again sets off thermals which carry up more moisture to replenish the Sc layer. The whole system often goes through a cycle lasting from a couple of hours to nearly the whole day.

When thermals reach an inversion

Figure 63 shows an isolated cumulus cloud rising through relatively dry air to collide with an inversion aloft. In (a) the rising thermal has just reached the inversion and produced a hump in it. At the edges of the cloud the down currents make a small dip in the inversion. In (b) the dip has been intensified, resulting in a rift through which warm dry air from higher levels is carried down below the inversion. The effect of this is to introduce extra warmth into the convective layer. As time goes on the process results in a slow lifting of the inversion. In sunny anticyclonic weather one often finds that the inversion rises slowly, sometimes as much as 2,000 feet, during a spell of prolonged sunshine.

Hook clouds

(c) shows what happens when there is a wind shear. The horizontal arrows on the left show a stronger wind above than below. Now the top of the thermal is distorted when it rises through the layer of shear. The top of the cloud may develop a hooked shape. With such clouds one finds the rising air on the up-wind side, with strong sink beneath the hook. The inversion break develops on the down-wind side.

Flow patterns round small cumulus

Real thermals can adopt many different forms; they are not all as simple as that illustrated in fig. 56. Some thermals have several cores very close together; others are twisted and distorted, making it difficult for glider pilots to find the best lift.

Figure 64 shows three flow patterns deduced from measurements made from a tethered balloon. The heights in thousands of feet are shown at the side of the diagram. The small cu have been drawn in near the 4,000 feet level. The airflow is shown by the arrows. In these illustrations the wind comes from right to left (the reverse of that illustrated in fig. 63).

Notice that:

(a) the flow tends to form a pattern like the letter P. The rising currents are twisted over by the wind aloft;

(b) the up and down currents can be found very close together at the edge of the cloud;

(c) there may be eddies well below the cloud where the flow branches off in a different direction.

These eddies at the edge of thermals can mislead

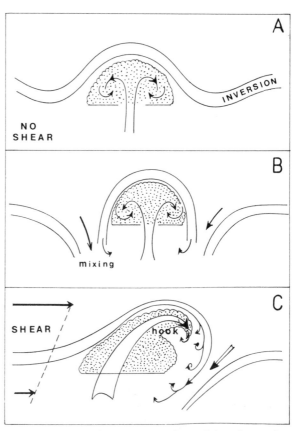

63. *Disruption of an inversion by a rising thermal*

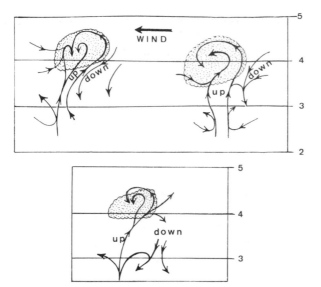

64. *Airflow round small Cu as observed by tethered balloon*

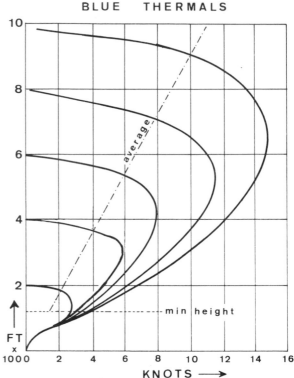

65. *Lift in cloudless thermals*

glider pilots whose variometers incorporate a 'total energy' system to compensate for changes of airspeed. If the aircraft encounters a horizontal gust there will be a temporary fluctuation in airspeed. This may be damped out by the internal gust filter; if it is not, the variometer will show a sudden change which is not related to any upward acceleration.

How fast the air rises in a thermal

Before the rising air can become organised into a soarable thermal, it usually needs to rise about 1,200 feet above level ground. (Mountain thermals are rather different.) Once the thermal has risen above the 1,200 feet level the flow usually develops the patterns similar to those illustrated in figs. 56 to 58.

Lift in cloudless thermals

Most cloudless thermals come to a halt when they encounter a stable layer aloft, although small ones die out earlier because they become diluted with air entrained from outside the thermal. On average the maximum rate of ascent depends on the height to which the thermal can rise. The deeper the convective layer, the stronger the thermal. This is illustrated in fig. 65. The rate of ascent usually starts rather slowly at low levels and builds up to a peak towards the

upper half of the thermal. There is then a rapid decrease towards the top. The curves neglect the rate of sink of a sailplane, but assume that the pilot was able to keep in the core of the thermal throughout.

Individual pilots find considerable variations between adjacent thermals even though conditions are apparently unchanged. The curves shown are an indication of the maximum rates of ascent in cloudless conditions. Average rates of climb may be only half these values. Pilots making fast cross-country gliding flights find it best to remain within the band of maximum lift and do not use the upper part of the thermal where the lift becomes weak.

Lift in cumulus clouds

Once a cumulus has formed there is an extra supply of energy available to strengthen the lift in a thermal. Figure 66 shows two graphs of the rate of ascent observed by radar tracking free balloons in cloud. Height is shown on the left and lift at the bottom of the graph. The pecked line shows the lift in moderate sized cumulus. This seems to reach a maximum some

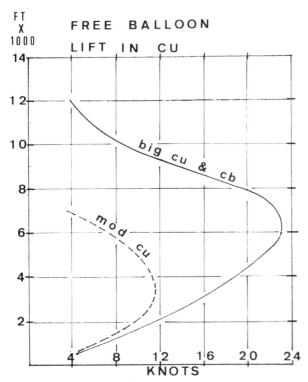

66. *Lift in cumulus clouds*

Forty knots is by no means the limit of updrafts. In severe storms the air may be rushing upwards at speeds of 100 knots; speeds like this can support large hailstones.

The decay of cumulus: holes in the clouds

In fig. 56 the thermal bubble was shown to have a mixing zone at the top where cooler and drier air was entrained into the cloud to dilute the thermal. This entrainment process can sometimes be observed if you watch the top of a Cu through a theodolite (or a pair of binoculars held rigid on a tripod so that cloud motion can be studied). Cu tops consist of numerous tiny domes and knobs which are constantly enfolding some of the outside air.

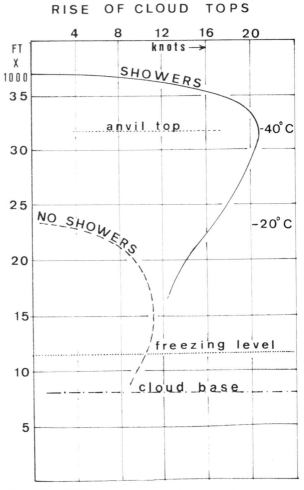

67. *Rise of Cb tops*

3,000 or 4,000 feet from the base; updrafts of nearly 12 knots were found.

With big Cu and Cb the maximum lift was found much higher up. In this case the maximum updraft of nearly 24 knots was found 6,000 or 7,000 feet above the base.

Figure 67 shows another method of deducing lift in clouds. In this diagram the rate of rise of the cloud top was observed by theodolite. Cloud tops are thought to rise at only half the speed of the maximum updraft within the cloud. As before, the height is shown on the left of the graph with speed along the horizontal axis. Two types of cloud were chosen. The pecked line is for clouds which did not produce any showers. The cloud base was high (about 8,000 feet) and the freezing level even higher so clouds could grow very high without giving showers. The 'no shower' cloud tops rose fastest through the 15,000 feet level where ascent rates of about 11 knots were observed.

The tops of Cb clouds which produced showers reached their maximum rate of ascent much higher. At levels around 30,000 feet the tops were seen to go up at over 20 knots. If one assumes that this speed was doubled inside the cloud then the maximum lift would have been 40 knots.

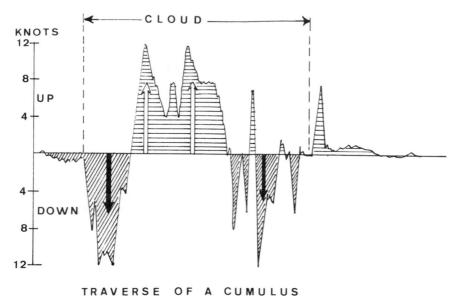

TRAVERSE OF A CUMULUS

68. Up and down currents when traversing a Cu

The entrainment process brings drier air into the cloud. This can result in pockets of evaporation. Evaporation produces a strong cooling effect. When enough dry air has been drawn into the cloud the cooling starts to produce holes in it. Evaporative cooling soon sets off strong downdrafts which sink through the middle of the cloud as well as round the edges.

It is difficult to see the early stages of decay when looking at a cloud from a mile or two away but one can sometimes see that the cloud shadow is no longer solid. With time the shadow grows more and more tattered. Little or nothing is left of the rising thermal at this stage but anyone flying into or under the cloud will probably find strong downcurrents. Holes can grow in any size of cloud, but when they develop in really big clouds the downdraft can become extremely strong. The phenomenon is more fully described in the section on 'downbursts and microbursts'.

Flight through a cumulus

Gliders which ascend in clouds try to remain in the main updraft where lift is often smooth until near the cloud top. Powered aircraft which fly straight through encounter a series of up and down currents. This is illustrated in fig. 68. The extent of the cloud is shown by the pecked lines while the strength of the up and down currents is indicated by the figures at the left.

Notice the strong downdraft at the left-hand edge of the cloud, the two main updrafts near the middle and the erratic pattern of lift and sink towards the right-hand edge. Some of the downdrafts are likely to be caused by the development of 'holes' within the cloud.

The complicated flow inside a cloud can be simulated in computer models. Figure 69 shows an example of the simulated flow in two large clouds extending up to 25,000 feet. Notice how some tracks turn round and start to descend well below the cloud top. Even the highest climbing air may be temporarily delayed by eddies.

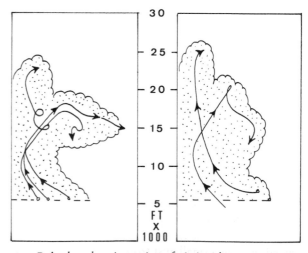

69. Calculated trajectories of air inside a growing Cu

6

The organisation of cumulus

The arrangement of cumulus sometimes looks haphazard but there are usually some rules governing the location of cloud. There are two separate processes which often combine to influence the clouds. The obvious factor is the ground surface. Dry surfaces heat up easily while damp areas remain cooler. Sunny slopes receive the most heating. Once the clouds have formed they begin to interact so that the pattern changes throughout the day. Downdrafts in between clouds may descend to the surface and trigger off fresh thermals nearby. A group of large clouds clustered together tend to monopolise the thermals and surround themselves with a region of sinking air which prevents isolated small clouds from growing anywhere near them. (This effect shows up best over an expanse of ocean where, in spite of uniform heating, the clouds show a wide range of sizes.)

Thermals leaving the surface

Figure 70 (a) shows an enlarged section of temperature and wind speed records on a sunny day when thermals were rising from the ground. The upper trace shows a slow rise of temperature followed by a sudden drop. At the same time the wind speed, which had been very low, suddenly began to be gusty. The cycle was repeated at intervals of 15–20 minutes. This fluctuation was the result of thermals breaking away from the surface and carrying the warmer air aloft. The gustiness was due to surface air being replaced by faster-moving air from higher levels.

Figure 70 (b) is a vertical cross-section through the unstable layer showing the kind of circulation which produces the series of gusts and lulls. As the thermal leaves the ground it is replaced by air descending from aloft producing an intermittent circulation. The descending air sometimes comes down with sufficient speed to form a small gust front which undercuts the warm surface air and triggers off another thermal.

Distribution of cumulus

Figure 71 shows the typical distribution of cumulus over fairly uniform level ground during the day. When thermals develop on fine mornings the first clouds are usually small and evenly distributed and the cloud base is fairly low. The life of each individual cloud is brief, only about ten minutes perhaps, but as one cloud disperses another forms close by.

By midday, when the temperature is several degrees higher, the cloud base has risen and the clouds are more widely spaced. The population of clouds has become less uniform. Some have grown larger but the number of little cumulus has decreased.

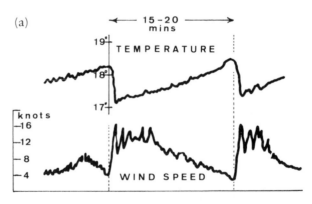

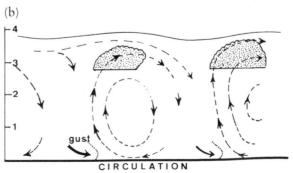

70. Thermals leaving the surface:
(a) temperature and wind speed traces
(b) circulation from ground to cloud

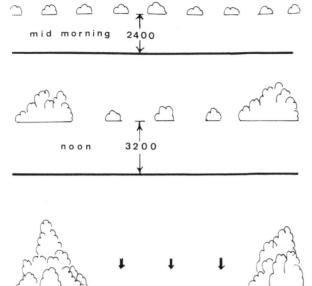

71. *Development of cumulus over flat land*

By mid-afternoon, when the surface temperature reaches its maximum, the cloud base is highest. The number of clouds has decreased further and those which remain are much larger, but there are wide gaps between individual clouds. In between the large clouds are areas of slowly sinking air which act to suppress the development of new clouds in the gaps. On average, the life of thermals and the space between them increases as the convection becomes deeper.

Figure 72 shows how cumulus develop over an undulating surface with wide, rather moist valleys and dry uplands. After a fine night the valleys are full of cold air and mist or fog may have developed beneath the inversion.

(a) After sunrise the temperature rises more rapidly over the dry uplands; the moister valleys are slow to warm up, especially if there is some early morning fog to be dispersed. As a result cumulus form first over the uplands, especially over any slopes facing the rising sun.

(b) It may take till nearly midday before the valleys warm up enough to set off thermals. When they do the cloud base tends to be lower there. On some days the cloud base over the high ground may be above the

cloud tops in the valley, even though the height difference is less than 1,000 feet.

(c) By early afternoon the air circulation has produced an almost uniform cloud base and there is little difference between hill and valley.

(d) Towards evening when the sun shines from a low angle thermals become very sparse. The western slopes of the high ground are then a good source of thermals and a few may be found over the valley as well, especially if cooling air starts to flow down the shaded east-facing slopes to disturb the remaining pockets of warm air still trapped in the valley.

Blue holes

On some days there are gaps extending for many miles between the cumulus clouds. Sailplane pilots refer to such cloud-free gaps as 'blue holes' and are reluctant to glide across them if the hole is very wide. Cloudless regions may indicate a total lack of thermals or even a region of slowly sinking air.

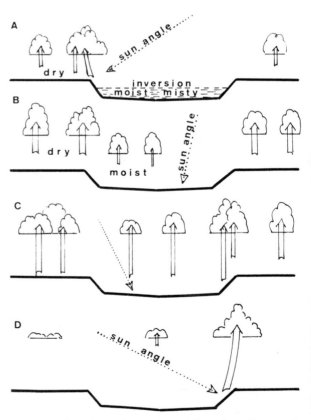

72. *Development of cumulus with wide valleys and high ground*

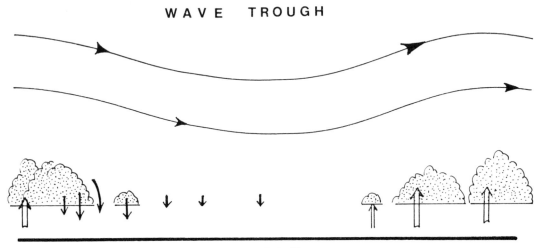

73. *A blue hole caused by a wave trough aloft*

Influence of waves

Figure 73 shows one possible reason. In certain conditions a series of standing waves develops in the airflow above the convective layer. These waves are often set off by a range of mountains far upwind; they are not always marked by any clouds. Convection is stimulated under regions where the upper wave flow is ascending but suppressed under wave troughs. As a result bands of cumulus tend to form and persist under wave crests and disperse under wave troughs. The individual clouds move across the region, growing on the upwind side and dissolving on the downwind side. In the cloudless gap there are no thermals and the air is often descending, making it difficult for a sailplane to glide across.

Influence of moist ground

In light winds a blue hole may be due to a difference in the surface heating. Figure 74 shows a wide and fertile valley filled with water meadows, woods and green crops. In such a region much of the sun's heat is used to evaporate the moisture from damp ground and growing vegetation. As a result there is little left to heat the ground and set off thermals. Cumulus clouds mark the thermals rising off the drier ground on either side of the valley but none appear over the moist ground. This effect is particularly marked where irrigation allows green crops to be farmed in otherwise arid regions. Desert thermals cease abruptly when the air passes over the cultivated region.

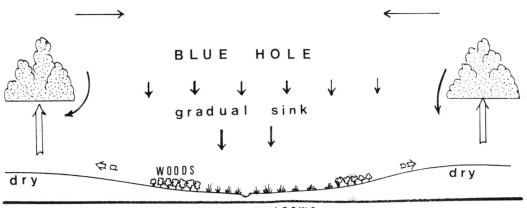

74. *A blue hole caused by a moist valley*

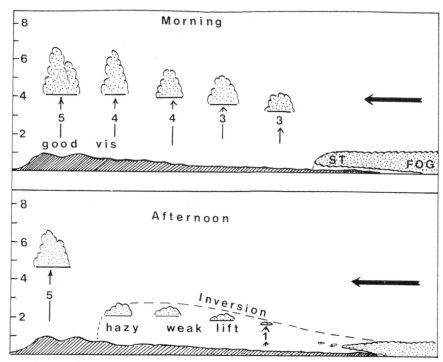

75. *Incursion of cold sea air spoiling convection*

Incursion of cold humid air from the sea

During the summer months, when temperatures are much lower over the sea than over land, the incursion of maritime air may stop or greatly reduce the depth of thermals. Figure 75 illustrates the change caused when an easterly wind carried air from the North Sea across the English Midlands to the Cotswolds. During the morning cumulus developed rapidly overland and tops reached about 7,000 feet by early afternoon. Over the North Sea the air was both cold and moist with a layer of sea fog below the inversion. During the day easterly winds carried this maritime air inland. The fog dispersed overland but the air never became warm enough for good convection. The boundary of the sea air was marked by a sudden drop in visibility. Within a couple of miles the large cumulus were replaced by scraps of stratocumulus and the strength of thermals decreased from 4–5 knots to barely 1 knot.

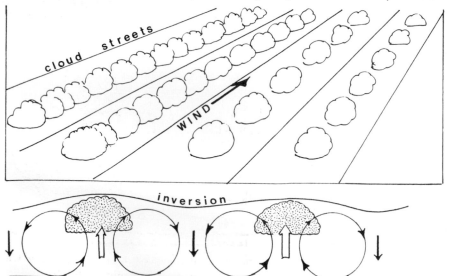

76. *Cumulus streets*

Cloud streets

Figure 76 is an example of cumulus streets. The upper part is a sketch showing the clouds arranged in lines lying almost parallel to the wind. Underneath is a cross-section illustrating the kind of air flow associated with streets. Air ascends beneath the cloud and descends in the gaps between streets. The air probably follows a helical path as it moves downwind.

Cumulus streets usually occur:

(a) when there is an inversion or stable layer which limits the tops of clouds

(b) when the wind speed is stronger than average and

(c) when the wind speed increases with height through the convective layer.

Alignment

Cloud streets are always aligned within a few degrees of the wind direction at their level.

Spacing

Streets are usually spaced 2.5 to 3 times the depth of the convective layer. Thus, if the cloud tops are restricted to about 5,000 feet, the streets are likely to be about 3 miles apart.

Oceanic cloud streets

Until satellite pictures became available it was not appreciated that very long cloud streets could develop over the sea as well as over land. These oceanic streets appear when cold air flows out over the warm sea in the rear of an eastward-moving depression. In the early stages the clouds do not grow very high and the streets are close together. As the air moves on to warmer regions the cloud tops grow higher and the spacing between the streets becomes wider. This happens by the decay and disappearance of some streets, not by the streets fanning out. When clouds become large enough to produce heavy showers the system of streets breaks up. Clouds then become arranged in large and irregular cells with clear spaces in between where air descends.

Cumulus lines

True streeting is marked by a large number of nearly parallel lines. High definition satellite pictures have revealed as many as ninety streets over England. Isolated lines of Cu, whose tops may vary considerably in height, are produced in different conditions:

(a) some cloud lines are formed over a mountain which is warmer than the surroundings. This sets off a series of thermals from the same spot. The clouds then blow downwind to form a line.

(b) a region where there is low-level convergence such as a minor trough or a sea breeze front also produces a line of cumulus.

(c) where there is inflow towards a very large storm cloud there may also be lines of cumulus feeding into the storm. Such lines are often curved and the clouds grow larger as the line approaches the storm.

Moisture and the persistence of cloud

The amount of cloud cover depends on the moisture in the atmosphere. When the air is very dry individual clouds tend to evaporate before they can grow large. This is because dry air is entrained into the growing cloud diluting the thermal and evaporating the droplets of water. If clouds develop in a closely-knit group, the inner clouds are protected from erosion and can grow much higher before encountering the dry air. This is illustrated in fig. 77. The isolated cloud on the left is soon eroded by its dry surroundings. The cloud within a group draws in moist air from its companions and grows much larger. Thus, a group of

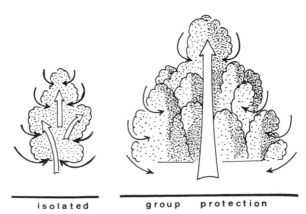

isolated   group   protection

77. *Variation in size between an isolated Cu and Cu protected by a surrounding group of clouds*

The edge of a cloud cluster over the Indian Ocean. Taken from just below 38,000 feet, it shows the north-east edge of a cloud cluster. The lines of small cumulus below are aligned with an easterly wind. The bigger Cu are in the middle distance, with Cb in the far distance producing the layer of anvil cirrus at the top of the picture. Notice the cloud-free zone at the bottom of the picture where the air is sinking and preventing any cumulus from growing

cumulus clouds can maintain itself much longer than an isolated cloud. The process is important for the development of cumulonimbus.

Very large clouds

The left-hand side of fig. 78 shows the shape of a temperature sounding made on a day of deep instability. The condensation level has been found by tracing the dry adiabatic from the surface temperature until it meets the line from the dew point. The shaded area above shows the energy which should be available to form cloud. The first clouds to form will seldom grow to this size. The sketch of a moderate-sized cloud on the right indicates the probable size of the early clouds. They only grow to the level marked 'mid-height'.

Each cloud carries moisture aloft and the air becomes progressively more humid. Finally, a full-sized cumulonimbus can develop, often from a group of large clouds. Large clouds generally contain strong upcurrents. The main body of the Cb tends to spread out when it reaches the top of the shaded section on the sounding to produce an anvil shape to the summit. The strongest part of the updraft overshoots the anvil level and pushes on into relatively warmer air above to form a dome. Very large and powerful Cb may develop domes penetrating several thousand feet into the stratosphere. In most cases the overshoot ends a few hundred feet above the anvil top. Flight above the tops of Cb is usually smooth because the air is stable there. However, a protruding dome can set off some turbulence several thousand feet above it. Such turbulence is generally only light in contrast to the interior of the cloud which is exceedingly rough.

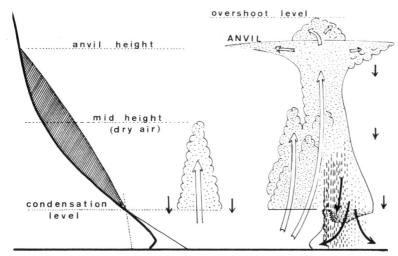

overshoot level

anvil height

ANVIL

mid height
(dry air)

condensation
level

78. Upper air sounding and the development of Cb

Looking for the lift associated with cumulus cloud

Sailplane pilots use cumulus clouds as signposts of lift. It generally takes many soaring hours to learn to read clouds successfully. The position of upcurrents is not always easy to spot and the cloud may pass through the growing stage and start to decay before one reaches it.

Figure 79 illustrates some of the different stages in the life of cumulus clouds.

(a) is the kind of very shallow cloud found just beneath a temperature inversion in fine weather. If the wind is light, the lift usually lies close to the centre. Sink is likely at the edges. Such clouds spread out under the inversion so that there is little or no useful lift in the cloud. A level, flat or very slightly concave base indicates that the cloud is still growing. Looking from the ground one may see the edges spreading out horizontally in several directions where the air comes up against the inversion. It is difficult to spot this when flying. The rate of climb tends to decrease rapidly when you get near cloud base. Although one may be able to reach cloud base it is often a waste of time to try.

(b) is a deeper cloud with more than one turret. Lift can usually be found while the cloud base remains firm and level. Down currents may develop between the turrets and penetrate well into the cloud. If the air outside is dry this penetration causes some evaporation and leads to decay.

(c) shows the decaying stage. There may still be some weakening lift just under the dome but the cloud base becomes ragged and fuzzy. One should avoid gliding under clouds like this. There is apt to be a lot of sinking air underneath.

(d) This sketch is of a large cloud growing in a wind shear. The wind is blowing from the left and is stronger aloft than below. The cloud consists of a series of thermals, the one on the left having formed last. Cloud tops look crisp and firm where the thermals are still active but they have begun to die out near the right-hand side. Here the ascending air is pulled over by the shear and starts to descend. The sinking side of the cloud looks fuzzy. One should avoid flying under this end.

(e) is a narrow column of cloud in shear. If there is strong lift in it the cloud will remain vertical, but as soon as the lift weakens the cloud is blown over by the stronger wind aloft. At the stage shown it would be wise to avoid flying under it; certainly one should avoid flying under the side to which it is leaning. The sink may be worse than expected.

(f) illustrates a large cloud with more than one active turret at the top. Even though the cloud base appears level it may be difficult to find just where the air is going in. Look for a few tendrils hanging underneath. These can be hard to spot when you are a long way below cloud base. The strongest lift is nearly always very close to these straggly bits, though not necessarily within them.

(g) is a more advanced stage than (f). Instead of tendrils there is a clear step in the cloud base. The strong lift will usually be found very close to the step under the higher cloud base. Stepped clouds like this are often found at a convergence line (such as a sea breeze front) where one air mass is moister. Even if there is no obvious zone of convergence this variation

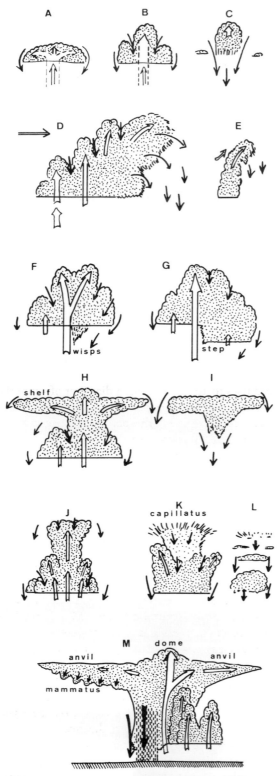

79. *(a) to (m) Some different stages in the life of cumulus clouds*

of cloud base gives a clue to the best lift. The cloud base is often a few hundred feet higher above any really strong thermals.

(h) and (i) are two stages in the development of a single cloud. (h) shows the active stage of a large cumulus growing beneath an inversion. The stronger lift penetrates a few hundred feet into the warm air aloft. A climb in this thermal will take you (temporarily) up into the clear air aloft. Most of the thermals are deflected by the inversion and spread out to form 'shelf clouds' which are mostly stratocumulus (Sc). The Sc layer cuts off the sun over a wide area and tends to stop any more thermal from forming. The remains of the Cu degenerate into a bedraggled tail hanging below the Sc layer.

(j), (k) and (l) form a series showing growth and decay. (j) is a multi-turreted cloud with several columns of lift in it extending well above the freezing level. Penetration of dry air into the central turret sometimes leads to a very rapid change. The top, which originally consisted of super-cooled water droplets, changes to ice crystals. The bulging domes of rising air turn into a hairy-looking mass called 'capillatus'. This may signify the shower stage has been reached. The fuzzy top often 'implodes' and collapses out of sight into the main body of the cloud leaving younger turrets to grow on either side. The third stage (l) shows the entire cloud falling to bits leaving broken layers at different levels. This region is full of sink.

(m) is a full-grown Cb cloud with a shaft of rain falling out of one side. When rain starts to fall the lift turns to sink and the air may begin to come down very rapidly. This can occur in one part of the cloud even though there is strong lift close by. It is possible to circle up in lift and find rain rattling off the canopy in one part of the circle. This is the moment to shift the circle before the whole region starts to sink.

A Cb like this may last for an hour or more with rain and heavy sink on one side and lift under newly formed cells on the other. A large anvil may grow outwards from the summit. If there is little shear of wind this anvil may extend in several directions. On the rear side of the shower these anvils often develop large downward sinking bulges called 'mamma'. Mammatus clouds are normally only seen after the shower has passed and they frequently indicate a large area of sink.

The next chapter deals with cumulonimbus in greater detail.

7

Cumulonimbus

The name comes from the Latin words 'cumulus' (a heap) and 'nimbus' (rainy cloud) but it covers a very wide range of conditions. The larger Cb clouds are potentially the most dangerous features of the atmosphere; they can produce violent up and down currents, destructive turbulence and very severe icing. The downdrafts may reach the ground with sufficient force to blow down trees and the outflow may produce squalls strong enough to blow over grounded aircraft. A Cb can generate hundreds of lightning strikes, cover the ground in inches of hail (some of it large enough to kill animals) and produce intensely destructive tornadoes. Clearly these clouds are worthy of study, preferably from a safe place.

When clouds produce rain

Cb are (initially) distinguished from other large cumulus by the fact that they produce rain. In order to produce rain it is usually necessary for the cloud to grow above the freezing level. Cloud droplets do not freeze as soon as they are cooled below 0°C; some may even remain liquid down to temperatures near −40°C. In regions of the cloud well above freezing level there is usually a mixture of ice crystals and supercooled water droplets. Where this occurs the ice particles grow at the expense of the water droplets. One may see this process acting in a domestic refrigerator. Anything moist tends to dry out; the liquid in it goes to build up the deposit of ice. In a cloud the process is fairly rapid, large crystals can grow in 10–30 minutes. These fall down into warmer air and melt into raindrops.

When the freezing level is very low, as in winter or when winds bring very cold air from polar regions, showers can develop from relatively shallow clouds. In contrast, cumulus which develop in warm tropical air usually need to rise well above 15,000 feet before a shower is produced. Some rain has been observed from clouds which did not reach the freezing level.

This is thought to be caused by growth of cloud droplets due to collision and coalescence. The larger droplets then start to fall faster and grow, gathering more droplets as they go. Eventually this process produces rain drops.

How fast rain drops fall

The microscopic droplets of water which form clouds descend extremely slowly. The fall speed only becomes significant when the drop size grows. For example, a drop of diameter 0.1 mm falls at 0.5 knots; when the diameter reaches 0.6 mm the fall speed increases to nearly 5 knots; if it grows to 1.4 mm the speed rises to 10 knots. With a diameter of 5 mm the fall speed is nearly 18 knots. If it grows any larger it is liable to break up due to the increasing airspeed. If the rain drop is carried aloft so that it freezes and turns into hail the fall speed increases as the hailstone grows. Some very large hailstones may fall at over 100 knots.

Where the energy comes from

The vast amount of energy within a Cb comes from the release of latent heat when condensation occurs. Figure 80 shows the difference between Cb in a cold unstable polar airstream and tropical Cb. In the left-hand section the freezing level is low so that the cloud can produce rain without rising to great heights. The tops of such clouds may reach to the base of the stratosphere, but in a cold air mass the tropopause is often rather low, perhaps 20,000–25,000 feet.

The tropical conditions are shown on the right. Here the air is much warmer, the freezing level is about 14,000 feet and the tropopause nearly 50,000 feet. The tropical Cb can rise to twice the height of a Cb in polar air, but height is not the only difference. Since warm air can hold far more water vapour than

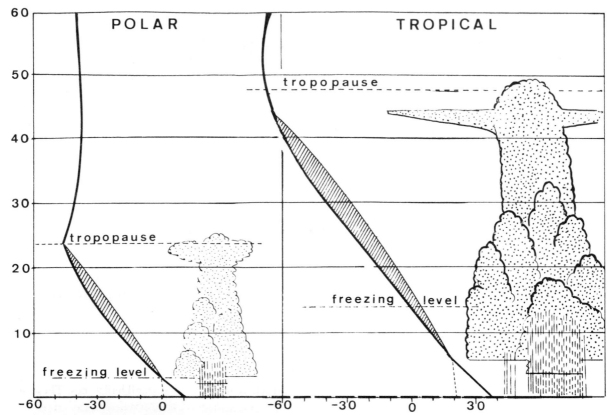

80. Cb in polar and tropical air; shape of temperature curves

cold air the tropical Cb release vastly more energy.

In temperate regions such as the British Isles and north-west Europe both types of Cb may develop. Cold air Cb occur throughout the year but the tropical monsters only develop during the summer months when very warm and humid air spreads up from the south.

Multicells and supercells: the effect of wind shear

The depth of the unstable layer is not the only feature which controls the development of Cb. The wind shear plays an important part too. If the wind velocity changes with height it is said to have shear. When the air is only slightly unstable a wind shear tends to distort weak thermals and disrupts the slowly-growing cumulus cloud. When the air is markedly unstable the cloud becomes too big to be blown over by the stronger winds aloft. Big clouds can deflect the stronger winds round the rising column or over the

top. As the Cb grows larger it starts to incorporate some of the faster moving, high-level flow into the circulation of the cloud.

The updraft from the surface is joined by a horizontal inflow of air from upper levels. This new influx is then diverted to form a massive downflow which lies near to (but does not hinder) the updraft. This powerful circulation forms one huge system called a 'supercell'. The important point is that the updraft is no longer checked by the rain-bearing downflow. A supercell differs from the more common kind of storm which is made up of a number of Cb cells clustered together.

Figures 81 and 82 illustrate the difference between a relatively small Cb which grows in an atmosphere with little wind shear, and the supercell which has a great shear of wind between top and bottom. Figure 81 shows the unsheared Cb. The arrows on the left represent the wind speed at various levels. In the first case the speeds are all the same. (a) shows the growing Cb. (b) shows it in the mature stage with an anvil growing out of the top and a heavy shower in the lower regions. The downpour of rain falls back through the updraft and eventually overwhelms it. (c) shows the cloud breaking up as the downdraft takes over. (d) shows the remnants of the original Cb. The downdraft has spread out on either side and set off fresh cells which will grow into Cb later.

Meanwhile, some of the cold dry air from aloft is drawn into the Cb and diverted to form the major part of the downdraft. The mass of precipitation, rain and hail, descends in this downdraft. The originally dry air cools as it evaporates some of the cloud moisture but soon becomes saturated with rain.

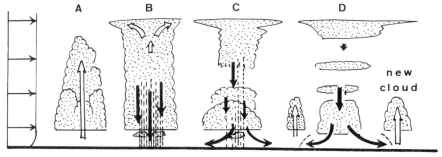

81. Cb developing and decaying with no wind shear

This is an example of a multicell storm. The new cells often become Cb before the original cell has collapsed so that the storm proceeds in a series of upsurges. Each cell moves with the general flow of wind but the new cells generally form on the right-hand side of the storm path. Thus as old cells die and new ones form the storm centre moves across the wind towards the right.

Cooling, added to the weight of rain, increases the descent speed.

Hail formation

Rain which develops in the sloping updraft is swept aloft where it freezes into small hailstones. These are carried forward where the updraft tilts so that they do

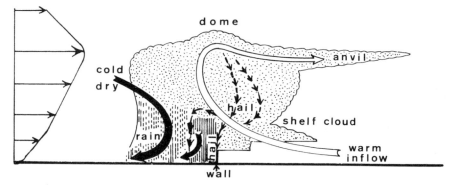

82. Cb in strong wind shear; the supercell

Figure 82 shows the basic structure of a supercell in simplified form. On the left the horizontal arrows show the wind speed increasing with height to the top of the storm and then decreasing at very high levels. The supercell has a sloping inflow drawing in very warm and moist air from the region ahead of the storm. This inflow turns vertical and ascends to the summit of the cloud forming an overshoot dome. It is then extracted by the strong upper winds and blown out ahead to form a long anvil of cirrus cloud extending for many miles.

not fall straight back into the ascending air. Many of the falling hailstones drop into warmer air and are caught up into the sloping part of the updraft. There they gather a fresh layer of water and are swept aloft again. Some stones may make many trips, collecting a fresh layer of ice on each circuit. Eventually they become too heavy for the updraft to support, or they fall to one side and miss it during the descending cycle. Then they enter the adjacent downdraft and hurtle down to the ground, the larger stones leading the pack.

The hail often arrives at the ground as a dense wall accompanied by a strong downdraft. Most of the downdraft lies beyond the wall of hail. As the wall advances the large stones give way to smaller stones and then to rain.

The essential feature of a supercell is the continued separation between updraft and downdraft. Once a supercell has become established it does not need a further supply of sunshine to keep the ground warm for new thermals. The system is self-stoking. It moves along like a giant vacuum cleaner sucking in the warm moist air from in front, releasing huge amounts of energy from the condensation, and blowing a big plume of exhaust out as an anvil of cirrus extending ahead of it. In tropical regions the anvil has been seen to extend for several hundred miles. The storm producing it can keep going right through the night, only dying out around dawn.

Figure 83 shows another version of a supercell with a more complex set of air currents. The updraft is shown branching into three, producing a short anvil on the upwind side and a much longer one blowing out downwind. The downdraft is shown spreading out underneath in several directions forming a cold outflow which undercuts the warm air. This outflow is marked by a gust front where the wind speed may rise to produce a squall, sometimes raising clouds of dust too. New Cb are triggered off by the outward spreading gust front. (The newest clouds have been shaded in on the diagram.) A point marked 'X' in this new cloud may develop into a 'downburst'. This feature is described further on.

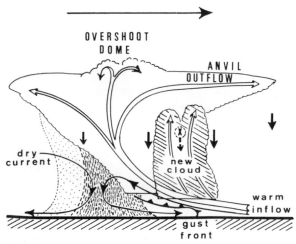

83. *Flow patterns in a Cb*

Figure 84 is a simplified 3-D sketch of the twisted air currents which form a supercell. The low-level current enters from the right (marked 'IN') and turns to rise vertically up to the summit dome before being carried away downwind by the strong upper flow, leaving by the arrow marked 'OUT'. The medium-level inflow enters the rear of the cloud much higher up from a very different angle. It twists above the updraft and then sweeps down to the ground and out. Where it strikes the ground there is an arc-shaped gust front which grows outwards accompanied by a sharp squall.

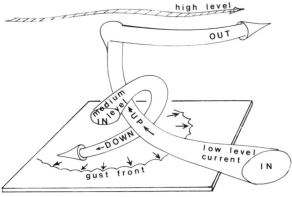

84. *Interacting currents in a supercell*

The thunderstorm outflow and squalls

The upper part of fig. 85 is a simplified cross-section showing the squall pushing out ahead. Eventually such a squall may extend far ahead of the Cb. The arrow on the right indicates a squall which has pushed many miles from the originating storm.

The lower plan view shows the spreading arc of the gust front marked as 'squall line' with a new Cb cell just starting.

Figure 86 shows a cross-section of a thunderstorm outflow. In desert regions these squalls can raise a great wall of dust known in the Sudan as a 'Haboob'. Similar dust storms occur in many areas of the world where there is sand and dust to be picked up by the squall. When the surface is firm, and the loose particles are held down by grass or other crops, the thunderstorm outflow may be invisible. This can make conditions very treacherous for pilots who are on their final approach to land before the squall arrives. Just before the squall arrives the windsock indicates a wind from the right. When the squall

*The curving arc of cloud produced by the outflow
from a tropical cumulonimbus. The original Cb had
decayed at this stage, leaving just this arc of cloud and
its associated gust line. The cloud was moving from
right to left (west to east in this picture)*

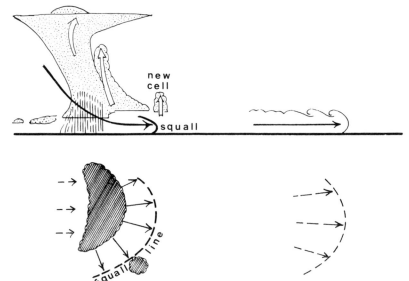

85. *Squall line and Cb outflow;
cross-section and plan*

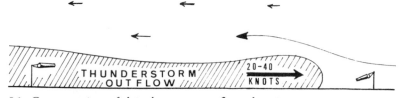

86. *Cross-section of thunderstorm outflow showing wind reversal*

arrives there is a sudden reversal of wind direction. An unsuspecting pilot descending into the outflow will find his airspeed suddenly falls as he comes down into a 20–40-knot tail wind during the last few hundred feet. Fatal accidents have been caused by this.

Figure 87 shows an enlarged view of a gust front. The cold outflow is one form of 'gravity current' and a similar structure is found with sea breezes. The outflow travels faster than the gust front and forms a nose where the cold air rises up and turns over. The warm air ahead of it is forced up over the nose and may form an arc of cloud. The cold outflow rises to a head, which may be a couple of thousand feet high but can be much less. Then there is a curl over where the flow breaks down into wake turbulence. The strongest part of the outflow is just above the surface. Drag reduces its speed close to the ground. (This is indicated by the backward curving marks in the diagram.)

Downbursts and microbursts

Figure 88 shows a cross-section and plan view of a downburst. Downbursts are columns of very rapidly-descending air produced by Cb. They are a feature of certain powerful Cb in hot climates and are unlikely to develop in the small Cb which occur in temperate regions. The arrows on the left of the figure show the wind shear. The maximum winds are just below the tropopause (here a little below 45,000 feet). This supercell generated three successive downbursts marked, in order of occurrence, DB1, DB2 and DB3. DB1 is the oldest and has weakened. DB2 is at the mature stage. DB3 is the newest one. It appears to have formed where some of the very dry air aloft has been pulled into the body of the storm just downwind of the dome marking the peak of the updraft. When this very dry air enters, it evaporates some of the moisture; this takes away much heat and makes the descending air relatively cold. Being cold, the rate of

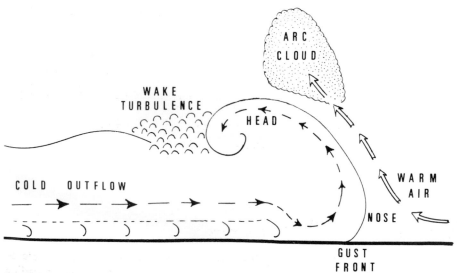

87. *Detailed cross-section of gust front and outflow*

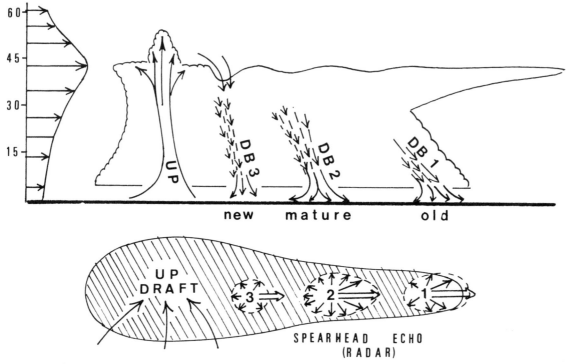

88. *Downbursts, cross-section and plan views*

descent increases until the downburst speed becomes very high.

The plan view of this supercell, which was observed by radar, formed a pattern called a 'spearhead' echo. There are other shapes of downburst clouds, not all of which have the spearhead shape. The plan view shows that the inflow came in from the southern flank before turning into the near vertical updraft. Underneath each downburst the air fans out in all directions when it meets the ground.

Figure 89 shows an enlarged section of a 'microburst'; this is just a smaller version of a downburst. The descending air sometimes forms a rotating column and when it strikes the ground it spreads out to form a sort of vortex ring marked by a swirl of dust. In the section shown between 10,000 feet and the surface the microburst descended at a calculated 50–70 knots and then produced a 60 knot horizontal gust at ground level.

The downburst may be accompanied by rain or occur in dry air below a Cb with high cloud base. In the latter case it may not be visible, which makes it more dangerous. Several accidents to airliners in the USA have been attributed to encounters with downbursts during approach to or climb out from airports.

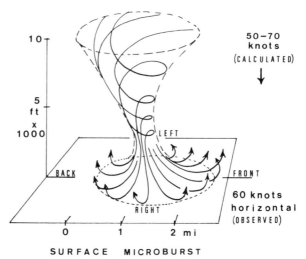

89. *Sketch of a microburst*

More Cb hazards

Hail damage

Cb which develop in a wind shear are liable to produce hail. Small hail will strip the paint off leading edges of an aircraft and may make considerable dents in nose cones. Large hail can cause more serious damage and may even kill animals and people out in the open. Hailstones weighing a pound have been found on a number of occasions in the USA and Europe and one exceptional stone which broke through the tiled roof of a house in Schleswig-Holstein weighed about 4.5 pounds. Many cars have been severely dented by large hailstones and a light aircraft (fortunately on the ground at the time) had the fabric on its wings and fuselage torn to shreds. Hailstones occasionally descend a long way from the cloud which produced them. This is one of many reasons for giving such clouds a wide berth if possible.

Lightning

Practically all lightning flashes come from Cb clouds which extend far above the freezing level. Here, supercooled droplets freeze, forming small hail (also called 'graupel'). When a droplet freezes, tiny splinters of ice may split off. These carry an electric charge of opposite polarity to the hail. The lighter particles are lifted by the powerful upcurrents, resulting in a positive charge accumulating near the cloud top with a negative charge lower down. The main centres of charge occur between temperatures of −10°C and −25°C, with a potential difference of up to 1,000 million volts.

Only about 20 per cent of flashes are from cloud to ground; most are between clouds and a few go vertically up into the stratosphere. Cloud-to-ground flashes start with a stepped leader branching downwards. This induces an upward stroke from the ground followed by a brilliant return stroke carrying a very high current, usually about 30,000 amps but occasionally as much as 200,000 amps. This instantly heats the narrow lightning channel to some 28,000°C, causing a shock wave which makes the thunder. Cloud-to-ground flashes typically have three strokes, but up to 26 have been counted.

Strikes on aircraft

Large metal aircraft provide good protection to the occupants, but compasses and sensitive electronic equipment may be upset and minor damage may be caused. The strike usually occurs on the tip of a wing, rudder or tailplane, where it leaves a small burn mark. Wooden and GRP aircraft can be seriously damaged or even destroyed by lightning. Recently a GRP sailplane (K-21), flying in clear air at 2,500 feet, was struck on the starboard wing tip while still some 2–3 miles from the Cb. The skin exploded outwards, the wings came off and the fuselage fell in two sections held together only by the rudder cables. Both pilots came down by parachute.

Distant strikes

Lightning strikes have occurred a long way from the main Cb. There is a severe risk to anyone launched by winch even when the Cb seems at a safe distance. 1,000 ft of steel launching cable makes a splendid lightning conductor. In recent years, at least three gliders have been struck during winch launches over southern England. In two cases the occupants escaped injury, but the third was fatal.

Most winches have a metal cage to protect the driver from injury by falling cables. This cage also protects them from lightning, but in one case the flash set fire to waste oil below the engine.

Tornadoes

A tornado is a violently rotating vortex formed beneath a Cb cloud where there is strong convergence of low-level air. Rapid rotation makes the internal pressure fall so that moisture condenses to form a funnel-shaped cloud extending down from the Cb.

Tornadoes have been seen in many countries but most have occurred in the USA and Australia, particularly during the approach of spring and summer. Some of the most destructive occur east of the Rockies, especially in Oklahoma and Kansas. Tornadoes are smaller and less frequent in NW Europe. However, England once had 105 separate events in a day. Near the British Isles, tornadoes are not confined to summer. They can also occur near cold fronts in the cool weather of autumn and early winter when the sea is still relatively warm but the air aloft is becoming very cold and unstable.

Tornadoes form under powerful Cb when there is a marked change of wind speed and direction with height. This wind shear makes the rising air rotate within the cloud, causing a fall of pressure known as a mesocyclone. Near ground level, the inflow of rising warm air is undercut by the cold downrush of air carrying rain and hail. A gust front forms with intense wind shear along it. Vorticity is concentrated by the air converging as it is sucked up into the Cb.

A 'wall cloud' appears under the Cb and a conical tornado vortex descends towards the ground, intensifying into a narrow tube. Before this touches down, loose debris and dust starts to swirl round above the ground. Tornado size varies from a diameter of 100 yards or less to as much as a mile. Wide tornadoes sometimes contain several smaller swirls known as 'suction vortices' which rotate round the main centre.

A tornado travels with its parent Cb, often at about 30 knots but occasionally as fast as 60 knots. The internal wind speed can exceed 300 knots; this is enough to lift railway trucks off the line, flatten strong buildings and hurl heavy steel girders deep into the ground. Debris sucked up into the storm has been found 55 miles away.

A single tornado generally lasts about quarter of an hour; some may be over in a few minutes while others persist for an hour. When it decays, the tornado narrows into a rope-like vortex which tends to lean over towards the horizontal before breaking and dispersing.

Waterspouts

These can form over a warm sea in a similar manner to tornadoes. A waterspout moving from sea to land has been seen to change into a tornado. Although they are more common over tropical seas, many waterspouts occur off the coasts of Europe and the UK. The gust front spreading out from a shower may act as a trigger by concentrating vorticity below the updraft of an adjacent Cb.

Dangers of flight in Cb

Cb clouds are not universally destructive. Many aircraft have flown through them without damage. Some glider pilots have climbed very rapidly to 30,000 feet and found the flight remarkably smooth until they left the core of the updraft. However, flying from updraft into downdraft has caused loss of control followed by structural damage due to overstressing the aircraft.

Fast climbs to high altitude increase the risk of hypoxia; icing can be severe, especially when leaving the lift. Gliders have had their controls frozen solid. Pilots who abandoned their gliders have been carried up under their parachutes and frozen solid. The pilot of a B-17, sent to investigate the dangers, met such thick snow at 30,000 feet that the intakes of all four engines were blocked. He had to glide out to lower altitudes before they could be restarted.

The tops of some Cb may extend up to jet-stream levels. One glider pilot was startled to find himself well out to sea when he emerged at the top.

Weaving through cumulonimbus

Pilots with long experience agree that it is unwise to try and penetrate an area of Cb clouds by following the valleys in between. These clouds can develop so rapidly that both the way ahead and the return to clearer weather can be blocked in a few minutes.

Problems for aircraft in circuit or on the ground

Watch out for a sudden wind change. As the storm approaches, the warm air first tends to flow towards it. Then a squall, starting under the downpour, spreads out ahead of the storm, reversing the wind direction with a powerful gust. This squall may be part of a microburst descending at 60 knots or more and spreading out horizontally. Downbursts have caused a number of accidents to passenger jets during the landing or take-off.

On the ground, light aircraft need to be pushed into a hangar or tied down securely. One gliding club had two parked gliders blown over in a few seconds as a Cb approached.

8
Waves

Whenever there is a wind blowing there are likely to be waves in the atmosphere. Often these waves are of such small amplitude that they are only detected by test groups measuring the glide ratio of sailplanes; some waves are huge and powerful, disturbing the air flow from ground level right up into the stratosphere.

Lee waves

One very important class of waves are those set off by the flow of air across a range of mountains; these are usually called 'lee waves', 'mountain waves' or

The top of a lenticular wave cloud at 22,000 ft above Strath Allen, Scotland. The cloud was aligned across the wind which was blowing from the NW (right to left)

sometimes 'standing waves'. The term standing wave is used because the wave remains stationary while the air passes through it.

Lee waves were first explored by soaring pilots in 1933 but very few pilots knew much about them until the late 1940s. Long after that there were still accidents attributed to pilot's lack of knowledge about the downdrafts which could develop behind a mountain range. Flight in lee waves is often so smooth that an aircraft can be flown 'hands off'. The smooth flow can be deceptive because it can exist close to areas of great turbulence at almost any level. A Canberra flying at 40,000 feet near Aberdeen one night was suddenly turned upside down by high level turbulence associated with strong waves set off by the Scottish mountains. In 1966 a Boeing 707 broke up in high-level turbulence to lee of Mount Fuji Yama. At low levels a number of aircraft have flown into mountain sides after being pushed down by the rapidly descending air of a lee wave.

Waves have also been found up to great heights. Lockheed U2 reconnaissance aircraft found waves above 60,000 feet. Nacreous ('mother of pearl') clouds which occasionally appear at about 80,000 feet are likely to be due to stratospheric wave. Reflective 'chaff' dispersed by a rocket at 100,000 feet was tracked by radar and found to have a wave motion.

very stable the flow tends to be along the valleys rather than over the ridges.

3. If the waves are to develop and extend downstream the wave energy needs to be kept within a definite channel, not dissipated by spreading out in all directions. Waves which are confined within a limited depth of the atmosphere are called 'trapped waves'. They occur when there is an inversion (or very stable layer) near the level of the ridge top, with a deep layer of lower stability above.

4. The wind speed should increase with height but the direction should remain fairly constant. A change of 20 degrees over a height band of 3,000 feet may not disrupt the wave system but a survey of wave climbs made by gliders over the United Kingdom showed that for 75% of all climbs of more than 10,000 feet the wind direction between top and bottom did not alter by more than 25 degrees.

Some basic requirements for lee waves

1. The low-level wind over the mountains should be at least 15 knots; speeds in excess of 20 knots may be necessary for bigger ranges.

2. The air must be forced to blow across the ridge line. For this to occur the wind usually has to be at right angles to the ridge, or within 30 degrees of this. Waves develop better if the ridge is long enough to prevent the air escaping round the end instead of flowing over the top. When lower levels of the air are

Cross-section of lee waves

Figure 90 shows an idealised cross-section of lee waves. The air flow is from left to right. The profiles of wind speed and temperature are shown on the left of the diagram. It shows the wind speed increasing with height. The temperature curve shows an inversion near the mountain top.

A cloud cap is shown over the crest of the ridge. Where the air descends the lee slope it warms up and evaporates the cloud. If you see a layer of cloud

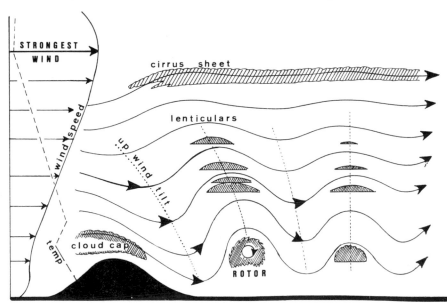

90. Cross-section of lee wave

apparently pouring over the ridge top but never reaching the bottom it is a reliable sign of lee wave flow. The cloud dissolves when the temperature rises during the descent. The break in the cloud is called a 'foehn gap' in Europe and sometimes a 'wave slot' in the United Kingdom.

On some days the flow down the lee slope may be much stronger than the flow upstream of the ridge, especially if the wave is very powerful. In extreme cases the downslope wind may be strong enough to overturn high-sided vehicles.

When the air rises up the face of the first wave there is a sudden drop in the speed of the surface wind. The change may take place within a few hundred yards; at one place the wind blows with gale force, a short way away it is nearly calm.

Rotors

When waves grow to a large amplitude the stream-lines rise and descend at a steep angle. There is often a region where the air is turning over and over under the crest of the wave. This is called a 'rotor'. The overturning makes the air highly unstable and the region is extremely turbulent. Rotors usually form at about the same level as the upwind ridge top. The rotor can often be seen as a bar of ragged cloud lying parallel to the ridge but when the air is very dry there may be nothing to mark the position of the rotor. Underneath the rotor the surface wind becomes light and variable or even reversed in direction. Bands of alternately strong and light surface winds are an indication of waves aloft.

Lenticular clouds

Well above the level of any rotor the crests of lee waves are often marked by smooth, lens-shaped clouds called 'Lenticulars'. They form when air is lifted above its condensation level. The top of the cloud is often a good indication of the shape of the wave flow at that level. The base may be flat or even concave. When the humidity varies with height a stack of lenticular clouds forms like a pile of plates.

Lenticulars are not usually seen at low level. When there is no marked rotor cloud the position of the wave at low level is often marked by a bar of Cu or Sc cloud which grows to a peak under the wave crest.

Cirrus cloud

Some wave systems produce a large sheet of cirrus cloud which starts above the ridge and may extend several hundred miles downwind without any undulations to suggest waves. This may be partly because the ice crystals which form the cirrus do not disappear quickly in any downdraft as do the water droplets of lower clouds. It can also be due to a special feature of some high-level streamlines. These have a jump at the first wave but no corresponding dip beyond. Such sheets of wave cirrus regularly appear on satellite pictures. The leading edge generally lies over the highest part of the mountains while the lee edge can be seen to extend with time until there is a very wide expanse of cloud. A streamer of wave-formed cirrus from the Pennines has been seen to reach Germany.

Factors which modify the wave

Width of ridge

The size and shape of a ridge does not alter the lee wavelength but it does affect the amplitude. The largest amplitude is produced where the width of the ridge fits the wavelength. Figure 91(a) shows that with a very narrow ridge a long lee wave only has a small amplitude; (b) shows how much larger the amplitude is when the ridge is the correct width; (c) shows a ridge which is too wide, and the wave amplitude is small. However, if the wind speed were to increase so that the lee wave became longer, ridge C would produce a much larger amplitude.

The lee wavelength changes with time, especially if fronts are approaching or moving away. The wave amplitude will then vary as the wavelength comes in or out of tune with the ridge width. A high ridge will produce waves of larger amplitude, always provided that the wavelength continues to fit the ridge width.

Successive ridges may be in or out of phase

When the air flow passes over two ridges exactly one wavelength apart the second ridge will produce a much bigger amplitude wave than the first (fig. 91(d)). However, if the separation between ridges is one and a half wavelengths the second ridge will be exactly out of phase; the wave will meet rising ground just where the air should be descending. This usually cancels out the wave behind the second ridge, as in example (e).

Steepness of lee slope, separation and vortex shedding

The shape and steepness of the lee slope has more influence than the windward slope of a ridge. Figure 92 shows examples of different lee slopes.

(a) is a gradual slope. Here the ascending wave flow is at a small angle and the wave crest is further out from the ridge.

(b) is a steeper slope. The ascending wave rises at a steeper angle; this brings the crest of the primary wave closer to the lee slope.

(c) shows a cliff on the lee side. The descending airflow cannot follow the contours; there is a line where the flow separates leaving a gap which is filled by an eddy. The effect is to smooth out the slope and the resulting wave has a shallower angle.

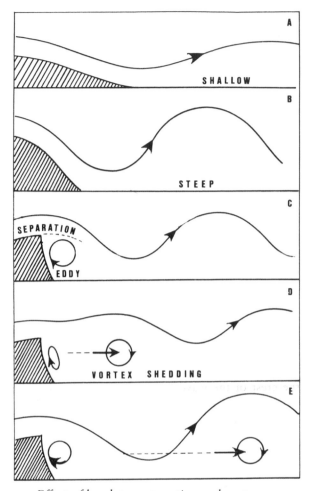

92. Effect of lee slope: separation and vortex shedding

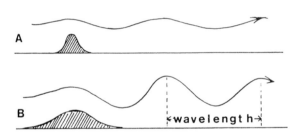

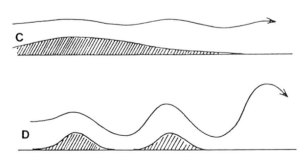

91. Ridge size and spacing; effect on wave amplitude

(d) shows how the lee eddy beneath the cliff breaks away from time to time and moves off downwind. The moving eddy alters the low-level airflow and changes the position of the wave. As the old eddy moves away a new eddy forms under the cliff. The process is termed 'vortex shedding'. When eddies are shed like this the wave crest first moves downwind when the original eddy is detached but then jumps back upwind.

(e) shows the wave when the old eddy is well out from the cliff and a new eddy has formed there.

Wind speed and wavelength

The lee wavelength depends to a large extent on the wind speed—the stronger the wind the longer the wavelength. There is a very rough and ready empirical rule relating the wavelength to the wind speed in the layer where waves are formed:

Wavelength (in miles) = 0.17 × speed (knots) − 1.6

20 knots = 1.8 miles	30 knots = 3.5 miles
40 knots = 5.2 miles	50 knots = 6.9 miles
60 knots = 8.6 miles	70 knots = 10.3 miles
80 knots = 12.0 miles	90 knots = 13.7 miles

To make a better calculation needs detailed knowledge of the structure of winds and temperatures aloft and usually requires at least a programmable pocket calculator with a large memory store. Accurate calculations can only be made with the aid of a computer.

High-level turbulence where flow reverses

Wave flow develops best when the wind speed increases with height. There are days when there is wave flow at lower levels but higher up the wind suddenly drops off or even reverses direction. This can make the tops of the waves break up to produce a very rough ride for anyone passing through the layer. Figure 93 shows the wind profile on the left with streamlines on the right. The speed is indicated by the length of the wind arrows. In the layer with smooth wave flow the wind speed increases with height, then there is a sudden check to the profile and the wind direction is reversed above. This is where turbulence begins.

Over the United Kingdom this wind structure is more likely when there are easterly winds at low level but westerlies higher up. When the low-level winds

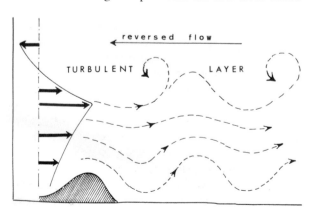

93. *Reversal of flow aloft and turbulence*

are from a westerly point it is very rare to find easterlies above them.

Phase lines and the tilt of waves

A line drawn through the crest or trough of each streamline is called a 'phase line'. When the phase line is vertical each streamline has its crest vertically above the ones below (fig. 94). This is often an indication of a stable wave which remains over the same region for several hours. The lift varies in a regular manner; it reaches a maximum several thousand feet above the ground and then decreases higher up where the wind speed is much stronger. In hilly areas the phase lines are often tilted into wind; the higher parts of the wave are further upwind. Far downwind of the mountains the phase lines are generally vertical.

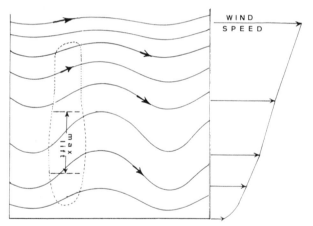

94. *Wave lift with vertical phase lines*

Some effects of tilt: asymmetry

A tilted phase line was illustrated over the mountain in fig. 90. The flow through a tilted wave is no longer symmetrical. It may develop unexpected complexities. Figure 95 shows computed streamlines with tilted phase lines and asymmetric wave flow. The dotted lines show two of the phase lines to emphasise the effect. A glider might start its climb at low level to lee of the ridge. As height was gained it would have to move upwind to stay in lift. At heights of 10,000 feet or more the lift might be upwind of the crest of the ridge. It has been known for gliders climbing near the top of one wave to be vertically above the descending edge of the next upwind wave. In these conditions wave clouds can be very misleading to a pilot at a different level.

Variations in wind speed through a wave

The asymmetry of the flow affects the horizontal wind speed. Where ascending flow is very steep the horizontal wind is very light. In 1970 a B-57F aircraft flying through a wave system at about 53,000 feet observed the indicated airspeed drop from 152 knots to 128 knots and then jump suddenly to 160 knots. (True air speeds would be more than twice these values at that height so that actual wind speed may have changed as much as 70 knots over a very short distance.)

A glider pilot climbing in near vertical streamlines notices very little drift and may (in exceptional circumstances) be able to gain height while circling. Such tactics would normally result in the glider being carried quickly back into the sink.

The lack of drift during a climb can easily deceive a pilot into supposing that the wind speed is universally light. Downstream from the wave crest the slope of the streamlines is shallow but the spacing is much tighter. This is a region where there is a strong horizontal wind as well as sink. If you turn downwind from the top of such a wave there is a rapid increase in ground speed due partly to deliberate speeding up in sink and partly to the increased wind. The combined effect may carry an unwary pilot straight through the next narrow zone of lift and out into the sink beyond.

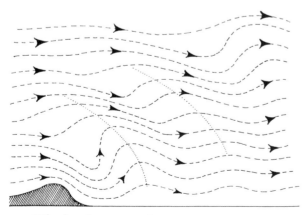

95. *Tilted and asymmetric waves*

Trying to go upwind may be difficult too because of the increase in head wind through the area of sink. It may be necessary to move along the wave bar to find a section where lift is weak before attempting the upwind jump. An inexperienced pilot can lose an astonishing amount of height trying to make the upwind transfer at the wrong point. One may feel ecstatic at 20,000 feet and, 5 minutes later, distinctly worried at 10,000 feet when the ASI is close to Vne, the variometer indicates maximum sink, and the next wave bar is still some distance ahead.

Vertical velocities in lee waves

The rate of ascent or descent in wave flow depends on the wind speed and the steepness of the streamlines. In many cases the streamlines have a very gentle slope and the rate of ascent in moderate waves is between 5 and 10 knots. In very strong wave systems speeds in excess of 25 knots have been found.

Variations in rates of climb

In simple wave the rate of climb usually varies in a regular way. The strongest lift usually occurs in or near the top of the stable zone (see fig. 96). When there are cumulus clouds this stable zone usually starts at the cloud top and extends up for two or three thousand feet. Above this level the lift usually starts to decrease. The decrease is often very gradual. As the wind speed increases the amplitude is reduced, waves tend to become flatter and the lift decreases. However, if there are very strong winds aloft there can be two different wavelengths, short waves at low levels and long waves high up. In these conditions lift may become stronger again during a climb.

Patterns of waves

The crests of lee waves are often marked by almost parallel bars of cloud. These bars can be seen on many high-resolution satellite pictures. Wave bars generally form parallel to the alignment of major ridges (fig. 97). They are not necessarily at right angles to the wind. The most consistent wave patterns seem to develop when:

(a) the stable layer is very deep (for example when there is an isothermal layer between 5,000 and 10,000 feet) and

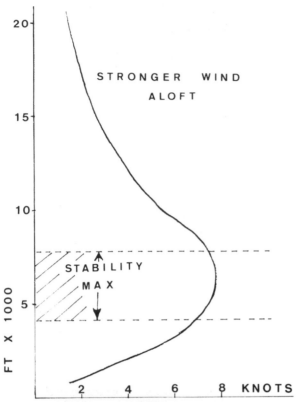

96. *Variation of lift with height in an average wave*

Wake waves from isolated peaks

Individual mountains which rise well above their surroundings set up a series of waves resembling the wake of a ship (fig. 98). These 'wake waves' can be seen in a sheet of stratocumulus when the cloud tops are not far above the peaks.

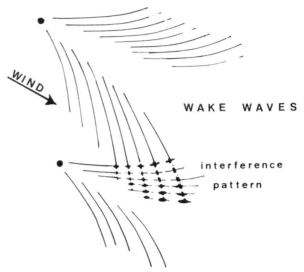

98. *Wake waves and interference pattern*

(b) there is a jet stream with speeds in excess of 100 knots at 30,000 feet on the low-pressure side of the wave area.

On such days wave clouds can start close to the west coast of Ireland, extend right across the United Kingdom and continue half-way across the North Sea towards Norway.

These wakes may be sharply angled back, like those from a fast-moving speedboat, or blunt and rounded like those from a tug plodding slowly along. The stronger the low-level wind, the more acute is the angle made by the wake waves.

When two or more peaks influence the flow their wake waves can cross, setting up an interference pattern. Where this happens the intersections of trough lines will disrupt the bars and produce broken lines of wave clouds. The combination of different patterns is often visible on satellite pictures but hard to detect at normal flying levels.

Travelling waves and billow clouds

There are a number of non-stationary waves which are not directly associated with the topography; they often produce a series of ripples in thin layers of cloud. Larger ripples produce a formation known as 'billow clouds'. Billow clouds have a very short wave length but otherwise look like a series of wave clouds moving over the ground. They form in a stable layer

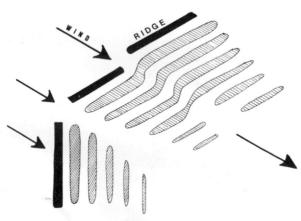

97. *Alignment of waves and ridges*

where there is a strong shear of wind speed in the vertical. The shearing effect tends to wrinkle up the layer into a series of small billows. The billows become steeper, start to curl over like breakers approaching a beach, and finally collapse into turbulence (fig. 99).

Clear air turbulence

The region of shear is not always marked by cloud but can be detected by high-powered radar because of variations in reflection at the layer. Radar and aircraft have been used in combination to study breaking billows. They found that clear air turbulence (CAT) occurs in this layer. Very small billows may also develop on the crests of much larger lee waves, usually near the upper limit of wave flow where there is a local increase of shear. They give rise to a 'cobblestone' type of turbulence.

The foehn effect

When air rises over a mountain range the moisture often condenses into cloud and then falls out as rain or drizzle. As the air descends on the lee side it is heated by compression and reaches the bottom both warmer and drier than it started. There are two main types of foehn wind depending chiefly on conditions upwind of the mountains. These are illustrated in the next two figures.

Figure 100(a) illustrates a moist airflow being carried up the windward slopes of the mountains. Heights are shown on the left and some temperatures marked on either side of the ridge. The air starts at 14°C, cloud forms at 5°C and over the summit the temperature falls to zero. Rain and snow fall out of the rising cloud which thus loses much of the moisture it started with. As a result, cloud evaporates on the lee side at a temperature of 1°C. Continued descent then raises the temperature at 3°C per thousand feet so that the lee valley is at 18°C, appreciably warmer than the windward side.

Figure 100(b) shows a different type of foehn wind. In this example the air on the windward side is so cold and stable that it will not rise up the mountain slope. The flow is said to be blocked. If the wind speed increases with height the higher level flow is diverted to sweep down the lee slopes. Since it starts out from

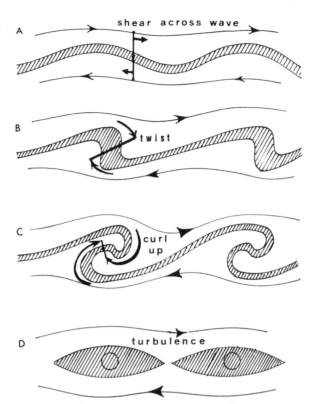

99. *Vertical wind shear distorting waves and producing turbulence*

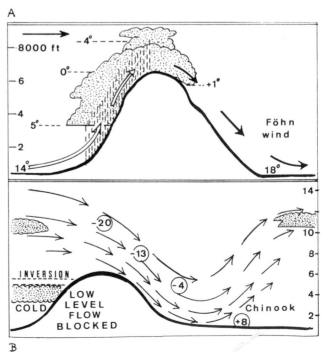

100. *(a) and (b) Two kinds of foehn winds*

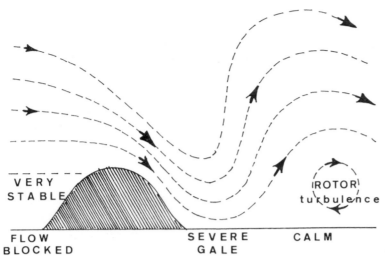

VERY
STABLE

FLOW
BLOCKED

SEVERE
GALE

CALM

ROTOR
turbulence

101. Severe downslope winds

high levels where the air held little moisture to start with the descent produces much more warming. This type of foehn wind occurs to lee of the Canadian Rockies where it is called a 'chinook'. The chinook is both warm and very dry; it can melt lying snow and evaporate much of the moisture too. The top of the lee wave is often marked by a high arch of cloud.

Severe downslope winds

With certain wavelengths the descending side of the foehn wind sometimes follows the slope of the mountains fairly closely. The air then sweeps down towards the plains with exceptional force. Surface winds have reached speeds of 100 knots on the lee side of the range, far stronger than over the ridge top. The pattern is illustrated in fig. 101. Notice the streamlines concentrating down the lee slope. The flow generally leaves the ground and rises up to produce a ferocious rotor cloud under which the surface wind falls light. Severe rotors like this are often set further back from the ridge than usual, are apt to be much straighter and

do not follow the undulations of the ridge line. In desert areas the wind is marked by clouds of dust rising from the ground.

These downslope winds have been extensively studied near Boulder, Colorado, but they are not confined to the large mountains of the USA. Even the smaller peaks of Scotland and the relatively low Pennine range can produce storm-force downslope winds. These are most severe when there is already a gale blowing across the country and conditions are also suitable for lee waves. The severity can take weather forecasters by surprise because it is more usual to find the worst of the gales are near exposed windward coasts, not sheltered areas well inland.

The force of severe downslope winds has blown down thousands of trees on the lee side of Scotland, torn off roofs and damaged buildings in Sheffield and caused the collapse of large, brick-built cooling towers at a power station. Less violent winds have blown over high-sided vehicles travelling along the A1 in Yorkshire. The striking feature of such winds is that the zone of damage can be quite narrow, with adjacent regions lying upwind and downwind not affected.

9

Some theoretical aspects of wave flow

The natural frequency of oscillation

The atmosphere is normally stable over most of its depth. The exception is that layer near the ground where thermals and cumulus clouds occur. Stable air tends to return to its original level after a disturbance. If it is given a jolt by passing over a ridge the air tends to oscillate up and down like a car with no dampers. The atmospheric oscillations have a natural frequency (called the Brunt-Vaisala frequency) which depends on the stability of the air.

The standard atmosphere, with its lapse rate of 6.5°C per km, has an oscillation period of about 590 seconds. For a big inversion where the temperature rises 5°C during an ascent of 1 km (3,281 feet) the period is reduced to 280 seconds. The greater the stability the more rapid is the oscillation. As the stability is reduced the period becomes longer until the point when the air is no longer stable. Unstable air has no restoring force and no natural period of oscillation.

The wind speed and wavelength

When waves are set off by a ridge they move away both up and down wind. The wave which moves down wind soon travels far away and vanishes but waves which move against the wind make much slower progress. Their speed depends on the wavelength; long waves travel faster than short waves. There is usually one set of waves whose speed is the same as the wind speed. When they move into wind they make no progress over the ground. They are then called 'standing waves'.

Their wavelength depends on the natural period of oscillation and the wind speed. If the wind speed was 10 metres per second, then the standard atmosphere (with its oscillation period of 590 seconds) would travel 5,900 metres during one full up and down oscillation. (Converted into other units a wind speed of 19 knots would give a wavelength of 3.2 nautical

miles.) In the case of the inversion the same wind would produce a wavelength of 2,800 metres (just over 1.5 nautical miles).

These values might be termed the 'natural wavelengths'. With marked stability and light winds these wavelengths are short. With little stability and a strong wind the wavelength is long.

The difference between the 'natural' wavelength and the 'lee' wavelength

In the real atmosphere there is a continuous variation in both wind speed and stability at different levels. Each layer of air lies between others which have different 'natural' wavelengths. The 'lee' wavelength is a compromise between the longest and shortest of the many natural wavelengths.

Trapping wave energy

When air passes over a ridge energy is radiated over a wide range of wavelengths. If lee waves are to develop fully it is necessary to trap the wave energy within a sort of atmospheric duct. Then, instead of escaping in all directions, the energy is retained to produce amplification at certain wavelengths.

Figure 102 shows three possibilities for a highly simplified air flow. In (a) the wind speed (shown on the left-hand side) is constant with height. On the right are rays indicating the direction in which energy at different wavelengths travels away from the ridge after the initial jolt. At the longest wavelengths most of the energy goes up at a steep angle while the shorter waves travel out at a flatter angle. As a result wave energy is lost upwards and little is left to be amplified downstream.

In (b) the wind speed is shown increasing with height. In this case the rays are bent forward and their

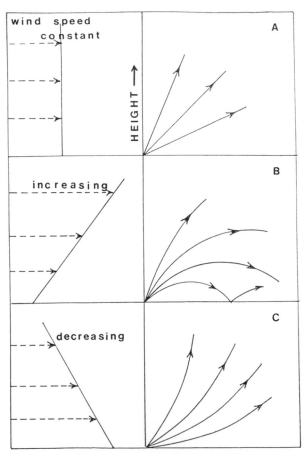

102. Effect of wind shear

out quickly; at very high levels the waves may develop large amplitudes but probably break down into turbulence in the stratosphere.

In (b) there is a very stable layer at low level and an increase of wind speed with height. The streamlines show a regular series of short waves at low level with a change to much longer waves at high level. This is an example of 'trapped waves'.

There is an intermediate stage known as a 'leaky' mode (in contrast to the 'trapped' mode). These waves grow steadily weaker downstream as energy leaks away. This is the commonest state of affairs. Strong waves occur just to lee of the main ridge but no wave train extends downstream.

Problems in computation

In spite of the ever-increasing sophistication of numerical models which have been introduced since waves were first studied, it is not always possible to calculate the streamlines of wave flow. There are several problems to be overcome.

The shape of the ridge

Nearly all the numerical models assume a smooth-shaped ridge which fits a convenient mathematical formula. It is supposed that the lowest streamline of flow follows the contours of this hypothetical ridge. All the rest of the lines depend on this, even the flow at high level, which is often very different from that below. If the low-level flow breaks away from the lee side of the ridges as indeterminate turbulence it is not possible to compute the streamlines. Satellite pictures show that the pattern of wave clouds does not necess-arily reflect the shape of the underlying mountains.

One cannot study wave flow in a wind tunnel because the space is too confined. It can be examined in a glass-sided water trough where a depth of a few feet of liquid is adequate for modelling flow which in the atmosphere takes up many miles. Water troughs show that the flow nearly always separates from the artificial ridges unless the wavelength fits the shape and size of the ridge. When the flow separates from the ridge turbulence develops and the shape of the wave above is affected. Turbulence is very difficult to compute and is usually handled statistically. This is not much use for calculating streamlines.

energy turned back to low levels. With strong enough winds aloft all the wave energy is 'trapped' within a sort of duct. After a while a resonant wavelength develops; waves of this length are amplified while others are damped out. When the winds at high level reach jet stream speed nearly all the wave energy is trapped beneath the jet and long trains of waves can extend far downstream.

In (c) the wind speed is shown decreasing with height. No waves are expected in this case. In the more complicated situation when the wind speed first increases with height and then decreases, any waves which develop low down break up into turbulence higher up. (An example of this was shown in Chapter Eight, in fig. 93.)

Figure 103 shows two examples of computed streamlines of lee wave flow. In (a) there is a constant lapse rate and no increase of wind with height. The waves are not 'trapped'. The energy spreads out upwards but decays down wind. Low-level waves die

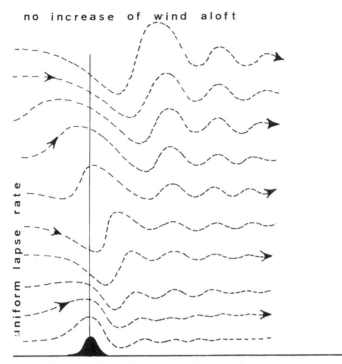

103. Cross-section of a calculated wave flow:
(a) with no increase of wind aloft and a uniform
lapse rate

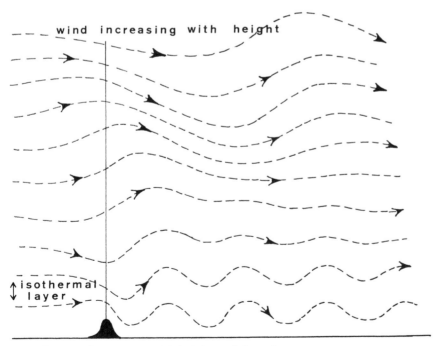

(b) when wind increases with height and there is an
isothermal layer above the ridge

Use of linear or non-linear equations

Early models employed linear methods. These may be very roughly described as models whose behaviour can be predicted from a relatively simple set of starting conditions. For example, the motion of a ball rolling down a smooth slope could be predicted by a linear equation. The effect of rolling a box of eggs down irregular steps is not so predictable. The box might be deformed at each bounce and the eggs would probably break at some stage. The end result seems likely to be messy, both mathematically and physically.

Unfortunately, when waves cease to be small undulations there are changes in the flow which are neither smooth nor accurately predictable by linear methods. With more computer power non-linear models were developed. These could model the change in the waves over periods of an hour or more. They have been used to study various kinds of 'feedback' when waves reach high up into the stratosphere to produce effects which are later reflected back to low levels. Some wave patterns look acceptably smooth for many minutes but then develop strange distortions and violent fluctuations of the streamlines. It seems easier to write a programme which can explain these anomalous patterns after the event than to predict them before they are observed in nature.

The need for three dimensions

The majority of mathematical models assume a two-dimensional atmosphere. The air is supposed to travel on a vertical plane cutting across the idealised ridge. This saves a lot of extra calculations which can only be made if the computer has a large enough memory. Three-dimensional effects have been modelled in recent years, but only for isolated peaks. They show the typical 'wake waves' angled back like the disturbance produced by a ship on the ocean. However, the effort needed to model the flow over the real irregularities of a rugged mountainous region is still too great to tackle.

10

Waves and cumulus

In the early days of wave exploration it was thought that wave flow ceased during the day when cumulus clouds formed. We now know that waves often exist above cumulus clouds but unless there are lenticulars to mark the wave pilots often remain ignorant of their presence.

Waves over isolated cumulus clouds

When a cumulus cloud grows it usually moves with the speed of the wind at the level where the thermal first formed. The air in a single thermal may weigh several thousand tons and it maintains its horizontal momentum as it ascends. The cloud still keeps its original speed even when it rises to a level where the wind is stronger. When this happens the faster-moving air aloft is deflected over and round the growing cumulus. This continues until the thermal is exhausted and the cloud starts to decay. For a time the growing cloud behaves rather like a hill. Waves develop where the faster moving air has to go over the top of the cumulus (fig. 104).

Sailplane pilots have found these waves by heading out into wind from a cloud climb, or sometimes after leaving the thermal at cloud base and steering into wind. The cumulus only acts like a hill while the thermal is still active inside it. Once the thermal stops the cloud starts to decay and ceases to deflect the upper winds. The wave then collapses and the cloud may begin to topple over in the stronger wind. The strongest sink occurs on the downwind side of the cloud which soon evaporates.

Waves parallel to cumulus streets

These may develop in two quite different situations.

1. When waves have formed overnight they control the development of cumulus during the morning. Thermals are inhibited under wave troughs but enhanced under wave crests. As a result, lines of cumulus develop parallel to the wave crests aloft and across the general wind direction. On these days the wind direction is the same up aloft as it is down at cumulus level.

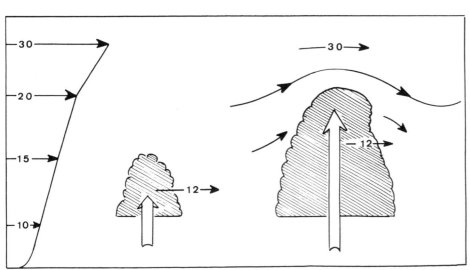

104. *Waves over an isolated Cu in a wind shear*

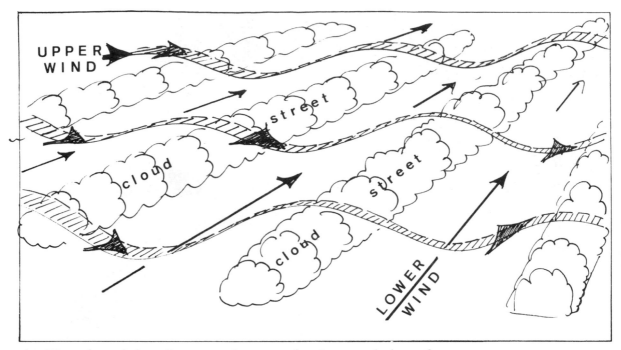

105. Waves formed above Cu streets when upper flow is across the lower flow

2. If conditions are favourable cloud streets form in lines parallel to the wind at their level. The cloud tops are limited by an undulating inversion which rises above the lines of Cu and sinks in the clear gaps between. These undulations act like a series of ridges and valleys to an air flow blowing across them. On days when the wind direction aloft is across instead of along the streets the undulating inversion sets off waves above the Cu (fig. 105). A street of clouds is much more effective than a single cumulus for producing waves.

Research flights by powered aircraft have found wavelengths varying from about 3 to 6 nautical miles with the maximum upcurrents of 4 to 6 knots above cloud streets. Some of the waves could still be detected at 29,000 feet (19,000 feet above the cloud tops). The presence of clouds is not essential for waves. When the air was too dry for clouds the streets were produced by regular lines of 'blue thermals' which also had waves above them.

Waves at right-angles to lines of cumulus

If the winds at low levels are fairly strong (25–35 knots) and there is an inversion limiting the height of cumulus, there are likely to be long cloud streets lying approximately parallel to the wind. Now if the wind direction above the inversion is unchanged and the speed continues to increase with height conditions become favourable for lee waves to develop. It is then possible to have long streets of cumulus capped by waves which lie at right-angles to them.

The conditions are illustrated in fig. 106. The left-hand side shows the profiles of temperature and wind speed. The temperature profile shows a well-marked inversion above the cumulus tops. The profile of wind speed shows a curving shape through the layer in which cumulus form and then a steady increase of speed with height above the inversion. The sketches show a plan view and vertical cross-section. The vertical cross-section shows the cumulus tops undulating in phase with the waves aloft, but generally maintaining an almost unbroken street. The plan view shows that the streets tend to thicken under wave crests and grow narrower under wave troughs. There are not always bars of lenticulars to show where the wave is located aloft; the air aloft is often too dry to form upper clouds.

Figure 107 is a pictorial view adapted from high-resolution satellite photographs. It shows an unstable airflow coming in across the coastline during the

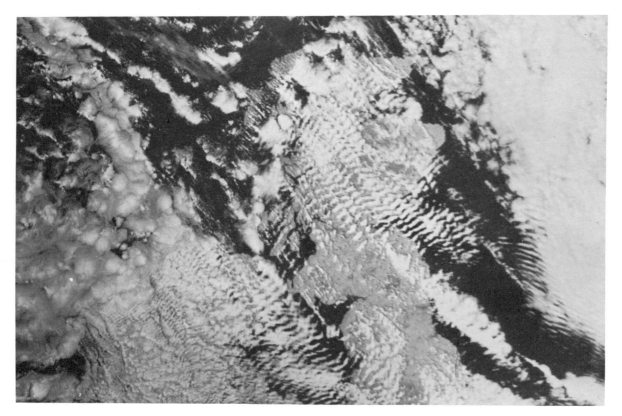

*Cloud streets and waves in a north-westerly flow.
Cumulus streets are visible over the Outer Hebrides
and the south west of Scotland. The wave pattern
begins over the western edge of the Highlands. The
bulbous cloud tops near the north west of Ireland are
stratocumulus formed by spreading out of cumulus*
(Picture from Dundee University)

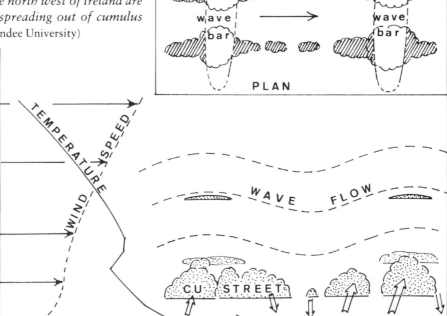

106. *Waves at right-
angles to Cu streets when
upper and lower winds
have the same direction;
plan and cross-section*

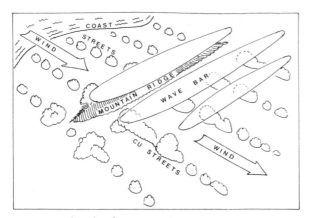

107. 3-D sketch of waves across Cu streets

morning. Cloud streets form soon after the air has reached the warmer land. A mountain ridge lying across the air flow sets off lee waves, and lenticulars appear high above the cloud streets. Well down wind from the ridge the wave bars die out but the cumulus streets extend much further.

Waves enhanced by cumulus over mountains

When there is deep instability lee waves do not readily develop over mountains until convection begins. Figure 108 shows an occasion when waves did not develop until cumulus began to form. The first cross-section, timed at 0600 hours, shows a nocturnal inversion filling the valley with cold stable air. This stable air reaches well up the side of the valley and only the higher peaks are above it. The wind above the inversion blows straight across as if the valley did not exist. At this time there is little or no hill lift on the windward slopes and no usable wave lift either.

The next section for 0900 hours shows the first cumulus developing over the mountains. Cloud forms here first partly because the sun-facing slopes warm up faster than the flat valley, and partly because the high ground is often drier. When the cumulus first develop, their roots are effectively attached to the mountains and any cloud carried away by the wind decays as it moves out over the valley. These mountain cumulus clouds form an obstruction to the flow aloft which then climbs over the cloud tops and drops down on the lee side. This initiates waves.

By 1100 hours the cumulus has grown large over the mountains and the wave flow has dipped down into the valley to produce a well-developed wave system. This is the best time for gliders to start wave climbs because the lift can be found on the upwind side of the growing cumulus clouds. These clouds can be soared just as if they were extensions of the hill.

By 1400 hours the convection has become so widespread that cumuli occur over both hills and valleys. At this stage it is very difficult to climb from the cumulus zone into the waves above. Gliders which are already high can remain in wave flow well above the cloud tops throughout the afternoon.

In the evening (1800 hours on the diagram) the convection has ended. The wave flow persists but without the extra boost from cumulus the wave lift is much weaker. Some of the decaying cumulus may be observed to change into lenticular clouds for a time.

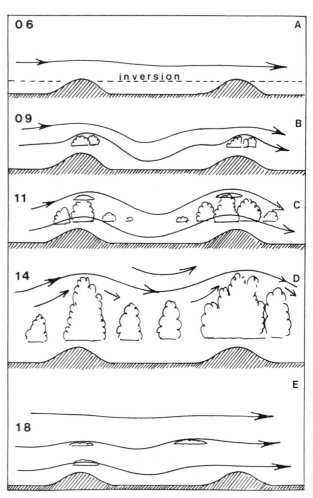

108. Waves and Cu over the mountains

The tops of wave cloud over the Grampians, east of Braemar. Notice the haze top marking the temperature inversion undulates over the wave troughs

The effect of showers on waves

Although it is usually extremely difficult for a glider to make contact with the waves above large cumulus clouds the arrival of a shower may produce a sudden change (fig. 109). Before the arrival of a shower the air below the cumulus tops is so disturbed by thermals that wave flow is non-existent. Heated sun-facing slopes may draw air up from the valleys and prevent any wave flow descending.

The passage of a big shower cuts off the sun for some time and the rain cools and wets the ground, making it difficult for more thermals to form. The passage of a large shower often clears away all the cumulus for a considerable distance. Now the air can sweep down the lee slope into the valley and bring waves down to low level. The change can occur rapidly. It opens a window for climbs to high levels at a time of day when convection currents would normally make it impossible for sailplanes to reach the wave.

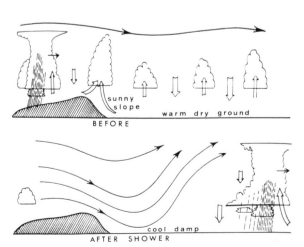

109. *Waves descending to surface after a heavy shower*

II

Flying in waves

When to expect wave

Waves are most likely when:

(a) the wind at 3,000 feet is forecast to be more than 15 knots (a mean speed of 25 knots is a useful average).

(b) the wind direction does not alter more than 25 degrees with height.

(c) the wind speed increases by 1–2 knots or more per thousand feet.

(d) there is an inversion or stable layer just above the level of the mountain tops. In the United Kingdom this often means the base of the stable layer lies between 3,000 and 7,000 feet.

The depth of the stable layer may vary widely between 1,000 and 7,000 feet. When the stable layer is very deep the waves are often widespread and the regularity of their pattern persists for long distances. If the stable layer is not well marked the wave pattern is apt to be irregular and the waves may not extend far down wind.

(e) there is a jet stream with winds of 100 knots or more at heights around 30,000 feet within 400 miles. The core of the jet should be on the low-pressure side of the wave area. For example, if the winds are from a westerly direction the jet stream should lie to the north of the mountains; with northerly winds the jet should lie to the east.

Figure 110 shows an upper air sounding made at Liverpool (Aughton) on a day in October when a sailplane reached 27,300 feet to lee of the mountains of North Wales. The temperature curve shows the stable layer extending up to about 10,000 feet. The figures at the right show the wind at different levels starting at 280 deg. 31 knots in the stable layer and increasing to 260 deg. 63 knots at 33,000 feet. Other sailplanes made high climbs over a region extending from South Wales to the Highlands of Scotland. There was a jet stream north of Scotland and an almost stationary front lying practically parallel to the isobars across central England.

Fronts

When a front lies almost parallel to the isobars waves are likely to occur on either side of the front, but the stronger waves are usually on the cold side (see fig. 111). When a front passes through, the lee wavelength alters. The change in wavelength may temporarily bring the waves in tune with the size and spacing of mountain ridges. When this happens the lee wave develops a much larger amplitude. A few

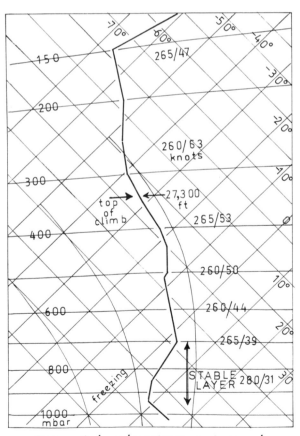

110. *Upper winds and temperatures on a good wave day*

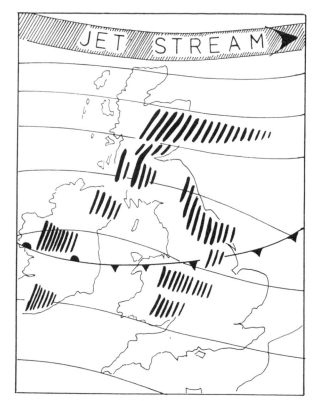

111. Wave pattern with jet stream and slow-moving front

high wave climbs have been made when a cold front crossed the area.

The problem on such days is that the ground may be completely obscured by cloud. This makes it difficult to locate one's position and consequently it becomes hazardous to descend through cloud. Satellite pictures often show well-developed lee waves above frontal cloud. These waves can seldom be exploited because the cloud base is too low and the depth of cloud too great for sailplanes to be launched safely.

Limitations of forecasts

Forecasts of lee waves are provided by various Met. Offices which regularly supply data for gliding, and also by main Met. Offices such as London Airport which issue warnings when strong lee waves are expected. These forecasts are based on smoothed data from upper air soundings; the data is entered on graphs which allow the forecaster to read off the lee wavelength; the level where waves will be strongest and the strength of the up-currents.

This graphical method assumes that the waves are 'trapped' (all the energy is confined to layers below the stratosphere) and hence the waves can extend a long way downwind of the mountains. The method fails for waves which are not 'trapped'. Although 'un-trapped' waves may only produce a single wave to lee of the ridge they can provide good height climbs at sites close to the mountains. However, cross-country flights are more difficult because no regular wave pattern extends down wind.

The graphical method failed to give a useful answer on the day when the United Kingdom height record for gliders was established over Scotland. (To be fair to the method the satellite picture showed that wave clouds did not extend far downwind of the Scottish Highlands that afternoon.) If a forecast office predicts lee waves with the level of maximum lift between 5,000 and 10,000 feet there is an excellent prospect of finding soarable lee waves. However, if the forecast fails to mention lee waves it does not always follow that waves will not develop. Gliders launched at flat sites are very unlikely to contact usable wave then, but unpredicted waves may be found at mountain sites.

Research units use a much more elaborate numerical method which requires a complex program to compute and plot the streamlines of wave flow. Such computations are usually carried out long after the event. This is not much use for an operational forecast unit.

Crossing lee waves in powered aircraft

The chief hazards are:

(a) loss of height when flying through the descending air. The vertical currents may exceed 2,500 feet per minute in a really strong wave, and rates of 1,000 feet per minute are not uncommon. Flight parallel to a lee wave may result in catastrophic loss of height in a few minutes. If such severe downcurrents are encountered the best course is to turn down wind until the ascending part of the wave is reached. This is usually safer than continuing to fly into a strong head wind until the sink stops. On some days the sink continues right up to the lee face of the ridge and the rate of descent can be far more than the aircraft's maximum climb in still air. Avoid flying on a cross-wind course. This merely keeps you in the sink for longer.

(b) Icing. Flight in cloud above the freezing level

may result in serious icing problems. Wave cloud can produce ice more rapidly than ordinary layer cloud.

(c) Turbulence. Flight in lee waves is usually exceptionally smooth but the smooth flow may exist very close to areas of severe turbulence. If the flight is made at low level to lee of the mountains the turbulence may be due to a rotor. Flight at high levels may meet turbulence when climbing or descending through a layer where there is a sudden change of wind direction. A reversal of wind direction with height can produce severe turbulence. Another source of turbulence is found in a region of strong vertical shear where the wind suddenly increases through a shallow layer. Such a wind shear produces a series of short waves which curl over and then break to produce patches of rough air. (An example was illustrated in fig. 99.)

Soaring lee waves in sailplanes

Lee waves enable sailplane pilots to climb to very high altitudes. Heights in excess of 30,000 feet have been reached to lee of hills which did not reach even 3,000 feet.

Starting the climb

This is often the most difficult stage because rates of climb are often very poor at low levels. There are some exceptions to this: on one occasion a pilot, who was being launched by winch, suffered a cable break when only a few hundred feet up. However, he was still able to climb away into the wave. On most occasions it is necessary to use lift from windward slopes or thermals to climb above the low-level turbulence into the smooth ascending air of a lee wave. Where there are large mountains a high aerotow may be needed to reach the lowest part of the wave.

Visual indications

On many days the low clouds do not, at first glance, appear to be due to wave. An active wave is often only marked by a slot of cloud-free air bounded by rather ragged cumulus or stratocu. From below the cloud seldom looks remotely like the text book illustrations of lenticulars. After climbing a few thousand feet it may be possible to see a lenticular cap to the low cumulus, or a smoothing out of the tops of stratocu, but by that stage the pilot is usually well established in the climb.

To start the climb one usually needs to be at the downwind edge of any clear slot in the cloud. There are often little fragments of cloud just upwind of the main cloud bank. These growing wisps mark the place where rising air first reaches condensation level. The best lift is usually very close to these fragments (fig. 112).

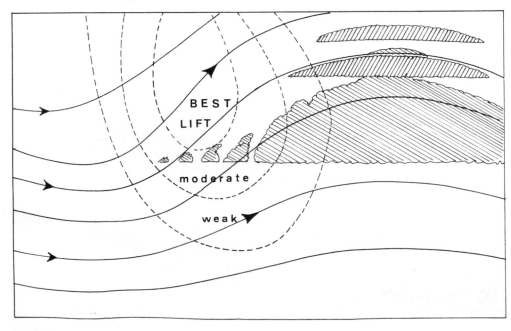

BEST LIFT

moderate

weak

112. Area of best lift at bottom of wave

Flight patterns

At low levels the wind speed is usually less than the speed for minimum sink. Therefore the glider is flown along the line of the wave cloud, usually a few hundred yards upwind of the leading edge of the cloud, as if soaring along a hill side. As height is gained it is often necessary to move further upwind since the wave front frequently tilts that way. At high levels the pilot may find he has crossed the clear slot marking the wave trough and is directly above the downwind edge of the next wave cloud.

Since the wind speed usually increases with height one often has to head more directly into wind as the climb progresses. Eventually one may be flying directly into wind and need to increase speed to stay in lift. On most occasions the wind speed does not seriously hinder modern GRP gliders which are designed to be efficient at speeds of 80–100 knots.

The difference between the indicated air speed (IAS) and true air speed (TAS) increases with height

At 850 mbar (4,781 feet) TAS = IAS × 1.07
At 700 mbar (9,882 feet) TAS = IAS × 1.16
At 500 mbar (18,289 feet) TAS = IAS × 1.33
At 400 mbar (23,574 feet) TAS = IAS × 1.46
At 300 mbar (30,065 feet) TAS = IAS × 1.64
At 200 mbar (38,662 feet) TAS = IAS × 1.95
At 150 mbar (44,647 feet) TAS = IAS × 2.26

Some years ago a large number of wave climbs of 10,000 feet or more over the British Isles were checked against the upper winds measured by radar tracking of balloons. The majority of these climbs did not encounter winds of more than 80 knots; only one reached a level where the radar found wind speeds of just over 100 knots. When speeds were corrected for altitude the peak value was about 80 knots IAS, and the majority were below 60 knots. The chief risk lies in exceeding the speed at which wing flutter may begin. This speed depends on TAS, not IAS which reads much less at height.

Searching for the best lift

Even when there is a well-marked line of wave cloud it is often necessary to try different locations for the best lift. The first step is to head further into wind. When the lift starts to decrease, slowly begin to turn across wind until lift reaches a maximum and then hold this heading. Try and find a feature on the ground to use as a marker for the best lift. If lift starts to decrease, turn into wind again and repeat the process. If this fails you may have come to the end of the wave; turn back and work along the wave in the other direction. Sometimes the good lift is over such a short beat that the best tactic is to make a series of slow 'S' turns.

When you reach levels where the wind speed is above the speed for minimum sink, head into wind

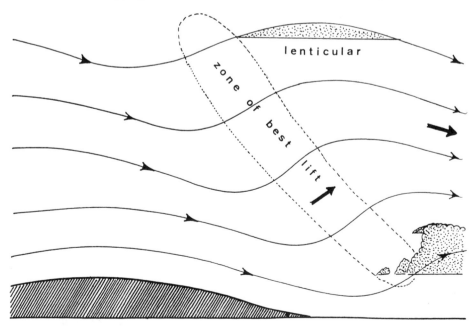

113. How zone of best lift tilts forward in some waves

and increase speed until you begin to make slow progress forward (you can check this against the cloud or some ground feature). Continue until the lift starts to fall off, then turn again, but only 40–50 degrees, and let the wind drift you back into better lift. As soon as the lift decreases turn directly into wind.

Figure 113 illustrates a pattern of streamlines which is often found among the mountains. Here the wave front tilts forward into wind. The climb starts from just in front of the cloud marking the wave bar. As you climb higher the best lift is found further upwind. Eventually at high levels the lift lies over the descending air low down. This kind of tilt also means that if you are high above the top of the wave bar there may be sink where lift is expected. If you push forward to the next wave there may be no lift until well beyond the cloud beneath you.

Looking at the leading edge of the cloud

The appearance of the leading edge of the cloud may give a guide to the slope of the streamlines. When the top of the cloud slopes backwards, as in fig. 114(a), the streamlines of the wave are likely to have gentle slopes. This is the commonest kind of wave. On a few occasions the front of the wave cloud may form an almost vertical wall several thousand feet high, as illustrated in fig. 114(b). This wall may be topped by an overhanging arch of cloud leaning into wind. This rather uncommon appearance marks an almost vertical wave front; here the streamlines are so close to the

The almost vertical face of a wave cloud to lee of the Ochils near Dollar. The wind was blowing from the left and the hook of cloud at the top curled over into wind, marking a region of turbulent air in a contra-rotating rotor

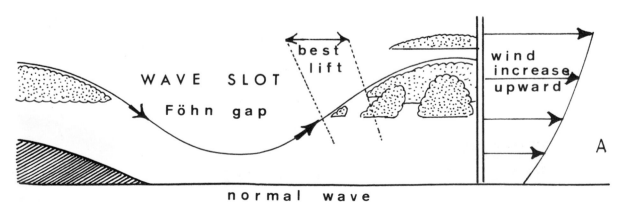

114. (a) Normal wave from gentle lee slope with wind speed increasing upwards

vertical that one can circle upwards as if in a thermal, or fly along the face of the cloud without experiencing any drift.

Moving along the wave

When there is a wave bar many miles long there are usually variations of lift along it. Some sections where the hills are higher, or slope down more steeply, may produce much better lift. It is often worth concentrating on this section if gain of height is the chief object of the flight. Wave lift is seldom constant, it varies from time to time so that patient exploration of the best area may reward you with a climb several thousand feet higher than elsewhere.

Most waves are aligned almost parallel to the upwind ridge. If the ridge has a bend in it the wave front often bends in sympathy. An irregular set of ridges may produce lines of wave clouds which are almost at right angles. By selecting one of these it is possible to make progress upwind with little or no loss of height.

Some points about high climbs

(a) **Oxygen** The most important item is adequate oxygen equipment. It is prudent to start breathing a controlled oxygen/air mixture by 10,000 feet. Many people believe they are still behaving normally at much higher altitudes but pressure chamber tests usually show that this is a dangerously false impression. Some years ago a Libelle pilot flying from Aboyne ran out of oxygen at some level far above 15,000 feet. The first stages of hypoxia produced great talkativeness on the radio. Then there was a silence. The pilot woke up very much lower down; the canopy had been shattered and the wings showed stress marks but fortunately the pilot managed to fly back to Aboyne.

(b) **Clothing** Really warm clothing is needed. The outside air temperature will fall far below freezing. Strong sunshine can maintain the temperature in the cockpit at a comfortable level for some time during a rapid climb, provided that the sun is shining in through the canopy. One beat along a wave may provide lots of warming sunshine, turning the other way brings more shade than sun and it can soon become very cold. If the flight is extended into the evening the cold can become almost unbearable.

(c) **Canopy icing** The moisture from one's breath can produce an opaque layer of ice inside the canopy. Trying to scrape this off is often a futile exercise because it reforms almost immediately. The extent of icing varies with different canopies. The first formation of ice can occur at the sides and the pilot may not appreciate that he has been flying in blinkers until the view ahead becomes fuzzy too. Eventually it may be necessary to open the clear vision panel, but this is apt to let in an icy draught. (Some canopies are designed to be opened a few millimetres in flight; this allows an inrush of drier air which clears most of the internal ice.)

(d) **Airframe icing** Many climbs can be done in clear air where airframe icing is no problem. It is wise

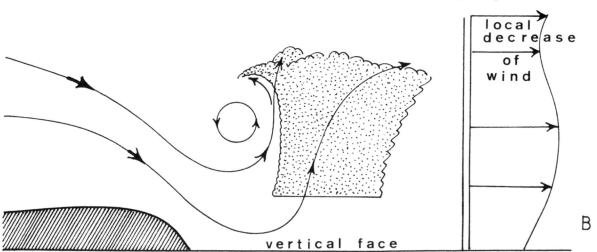

local decrease of wind

vertical face

B

(b) Vertical face to wave with steep lee slope and local decrease in wind speed aloft

to keep out of all cloud when above the freezing level. Flight in wave cloud can produce icing very rapidly indeed; even apparently thin lenticulars which look innocuous can soon cover both canopy and wings with ice. Difficulties arise if the wavelength changes and the lenticular which was previously behind suddenly moves forward to envelop you in cloud. Similar development of cloud occurs when a front is approaching.

(f) **The approach of fronts** When flying at high level the thickening frontal cloud can usually be seen many miles away. It is not always realised that extra moisture can be brought in by winds blowing almost parallel to the front you are watching. When this moist air reaches the mountains and rises in the wave system there is a rapid development of cloud all round; this may happen some distance ahead of the main frontal cloud. It is disconcerting to become surrounded by cloud much earlier than expected.

(g) **Sunset** The onset of darkness can catch out the unwary. It is much lighter up aloft and the deepening gloom at ground level may not be obvious to a pilot at 20,000 feet. In the summer months it never gets completely dark at high levels over Scotland, but the ground may be invisible.

Cross-country flight in waves

Going upwind

It is often wise to resist the temptation to push straight into wind from the top of a particularly high climb. If the lift has been very good at that spot then the sink will also be stronger when you head for the next wave. The best region for crossing is where the wave is relatively weak.

Sometimes the wave bars form a herringbone pattern where waves from different ridges intersect. It may then be possible to move from one wave bar to another aligned at a different angle without the enormous loss of height involved in a direct upwind dash. Where the two wave bars intersect you may find extra depth of cloud or a series of lenticulars stacked one above the other marking particularly good lift.

Signs of collapsing wave

Some wave systems hold their position and strength for many hours while others have a brief existence.

When the phase lines are vertical (i.e. the wave has no forward tilt) the pattern is often stable for several hours. Tilted waves are apt to be less steady. Waves which develop above cumulus are very likely to appear and vanish over short periods.

One indication of impending collapse is the appearance of tongues of cloud spreading across a previously clear gap in the cloud. It is not easy to distinguish between a general increase in moisture which simply fills in the gaps, and the strands of cloud which show that the trough of the wave is no longer deep enough to evaporate the cloud.

Watch for wave bars which move closer together. Gradual changes of wind speed or the inversion level may lead to collapse of the upwind wave. The associated cloud bar does not always dissolve immediately; instead it drifts off downwind and begins to close up on the next wave bar. This starts a sort of domino effect with all the waves dying out.

Deteriorating weather

It is possible to stay high in wave while a front passes right through the area. In the lee of some mountains a small slot remains open throughout, but this cannot be relied upon. As a front approaches there is often an increase of humidity at many levels. When cloud forms it does not only spread in from the direction of the approaching front. As the moister air reaches the mountains cloud can develop all round. It may even develop behind you while you are facing into the wind. Unless the best escape route away to low ground has already been planned the experience can be frightening. If the cloud has begun to thicken up at flying levels it is often only a matter of time before a layer of much lower cloud forms below the level of the mountain tops.

Waves above cumulus

There are a number of summer days when one can fly long cross-countries below cumulus without any idea that there is wave flow above. An unexpected variation in thermal strengths is often the only indication of wave higher up. Wave lift is likely where thermals become far stronger than the average for the day. If you encounter a series of clouds which, despite appearances, seem to lack good thermals, they may be under a wave trough. An unusually wide area of heavy sink between clouds is also a sign of a wave trough. (An exception is the normal band of sink between parallel streets of cumulus.)

Climbing from cumulus to wave

Successful transition from thermal to wave is often a matter of seizing the opportunity when it arrives. If a particularly strong thermal makes it worth going on into cloud one should aim to head out directly into wind from the top of the climb. If there is wave the exit into clear air will be smooth; there should be none of the heavy sink so often found just outside a cumulus. It is not necessary to enter cloud if one is already near the upwind edge. A quick move out into the clear from near cloud base may take you into the wave.

The first part of such a wave climb can tax the patience. Lift is often very weak indeed and it is essential to climb high enough to be above the next cumulus cloud when it arrives. This is why it is easier to become established in wave after a cloud climb. Lift generally improves above the cloud top. From above one can make out the wave pattern clearly, below cloud it all seems a haphazard grouping of cumulus.

Entering wave from cumulus streets

Unexpectedly prolonged sink below a street of Cu with occasional bursts of very strong and often rough thermals can be a sign of waves aloft. The wave is best entered from the upwind end of a cloud street. If you try a cloud climb half-way along the street and head out into a stretch where the tops are lower there is seldom time to climb clear of cloud before the next top arrives to break up the weak lift.

Is it worth using waves above cumulus?

If you enjoy being above the cloud tops that alone is worth the effort of getting up to the wave. If you are trying for a fast flight over fairly level country and are doing well below cloud it is probably better not to waste time in the tediously slow early stages of a wave climb. During one competition some of the pilots climbed high in wave while others stayed below; neither group seemed to have a particular advantage.

One big advantage of a wave climb is the opportunity of a long glide at the end of the day when all the cumulus are decaying on the route home.

12

Local winds

This chapter describes various winds which are not directly controlled by the pattern of isobars. These are winds which blow for periods which are too short for the coriolis force to have its full effect. The airflow is often across rather than parallel to the isobars and is controlled chiefly by differences in surface heating. Coastlines, mountains and valleys have more effect than isobars on these winds.

Sea breezes and the sea-breeze front

Sea breezes develop on fairly calm days when sunshine makes the air much warmer inland than out over the sea. The various stages are as illustrated in fig. 115.

(a) Sunshine raises the temperature overland and thermals carry the heat upwards for several thousand feet. Heating makes the air expand and tilts the isobaric surfaces so that they begin to slope downwards towards the cooler regions over the sea. The air aloft then starts to flow down this slope. This process reduces the total amount of air over the warm land. The reduction is very small, about 0.2% or enough to bring the pressure down by a couple of mbar, sometimes more. In summer this fall of pressure can produce a heat low deep enough to show up on weather maps.

(b) At low levels the cooler air from the sea then starts to flow inland towards the low pressure region. This is the start of the sea breeze. There is usually a boundary between the cool, relatively moist, air from the sea and the warmer land air. This boundary is rather like a shallow cold front. It is called the 'Sea Breeze Front' and as it pushes inland a band of rising air is concentrated along its length, providing lift for sailplanes and also for a number of birds and insects.

(c) The heating which produced the sea breeze decreases during the late afternoon and ceases before sunset, but the front continues to move inland for several hours more as a weak feature.

(d) Some time after midnight, when the land becomes colder than the sea, the process is reversed. Pressure rises slightly inland as the heat low fills and a shallow flow from land to sea develops in the lowest layers of the atmosphere. This is the land breeze; it is almost always a much shallower and weaker feature than the sea breeze.

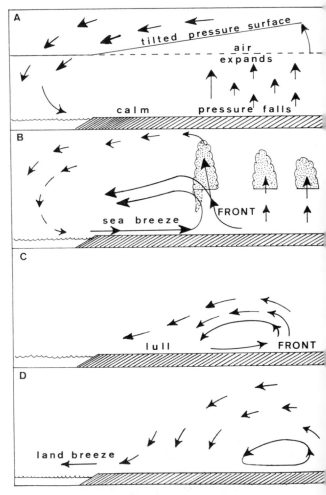

115. Development of sea breeze circulation

100

Factors which influence the sea breeze

The sea breeze is a good example of a phenomenon which is affected by a number of factors, some of which act in opposition. The main factors are:

Heating

i.e. the amount of heating which occurs inland. The increase in solar heating during the summer months strengthens the sea breeze. During a dry spell the land heats up faster and the sea breeze becomes stronger; after a wet spell, when a larger proportion of the solar energy is spent in evaporating surface moisture, the sea breeze is weakened.

Land/sea temperatures

The sea temperatures are very slow to change but overland the temperature can fluctuate rapidly. During the spring and early summer the sea is always relatively cold but the land can become hot during a sunny spell. This contrast helps strengthen the sea breeze. The sea slowly warms up during the summer so that by autumn it is often warmer than the land, which no longer receives so much heat from the sun. As a result sea breezes become weaker in autumn and are very rare in winter.

Near the coasts of the British Isles and north-west Europe the first few sea breezes appear on sunny days in March; then the numbers increase month by month and reach a peak in June. From July to the middle of September the numbers show little decrease but in October they dwindle rapidly. In the winter months the land is too cold to produce a sea breeze.

Previous wind

The wind which exists before any sea breeze develops can help or hinder the penetration of the breeze during the day. An offshore wind delays the sea breeze and may hold it up indefinitely if it is too strong. An onshore wind helps the sea breeze to penetrate a long way but the constant supply of cool sea air makes it difficult for a heat low to develop inland. The sea-breeze front rarely appears if there is no opposing offshore wind. It is the conflict between the two winds which produces a front.

Depth of instability

The depth of inland convection modifies the strength and penetration of a sea breeze. If the air is very stable so that thermals are unable to reach 3,000 feet, even by mid-afternoon, the sea breeze circulation does not develop properly. The sea breeze then makes little progress inland. The best conditions for deep penetration seem to be a moderate depth of convection with cumulus up to 6,000–9,000 feet inland. Very deep convection tends to halt the sea breeze near the coast. Here the cloud may grow into Cb with showers concentrated along the line of the front.

Latitude

The latitude influences the sea breeze penetration too. There are two reasons for this. Obviously the heat from the sun is greater in tropical regions but this is not the only reason why tropical sea breezes can extend much further inland. The cool sea breeze starts off blowing directly towards the heat low inland but after a few hours the coriolis force deflects the flow so that the wind then blows round the heat low instead of into it.

The coriolis force is stronger in high latitudes (towards the poles) than in low latitudes nearer the equator. In the latitude of the British Isles the coriolis force changes the direction of the sea breeze in a few hours. Instead of blowing straight inland it is deflected to the right. In tropical regions the coriolis force is so small that the sea breeze is not significantly deflected; as a result it pushes straight inland very much further during the day.

Affect of hills

Steep hills (which rise to 1,000 feet or more) tend to delay the progress of the sea breeze and deflect it along valleys. Wide estuaries and broad valleys offer an easier passage. As a result tongues of sea air may penetrate far inland where the ground is level, spreading like the tide round higher ground to meet on the far side. The sea breeze can climb over gentle hills, especially when their slopes face the afternoon sun.

Shape of coastline

The effect of the sea breeze is also influenced by the shape of the coastline. On a narrow peninsular the sea air comes in from both coasts and forms a convergence line near the middle; this gives a temporary boost to convection. When there is already a flow of air across a peninsular the line of the sea-breeze front may be carried towards the lee coast and may be even found as a convergence line some distance offshore. A large bay extends the sea-breeze front over a wide arc and tends to reduce the strength of lift along the front.

Tidal effects

In some regions the arrival of the sea-breeze front is influenced by the time of high tide; this has been observed even in regions where the ridal range is not great.

Soaring sea-breeze fronts

Timing

The front seldom crosses the coast much before 10 o'clock (local time) and may be delayed until mid-afternoon if there is an offshore component to the wind. In calm conditions the front moves inland at between 5–10 knots (and may come to a halt at times) but the speed can be as much as 15 knots in some areas. This makes it unlikely that soarable sea-breeze fronts will be encountered much more than 50 miles from the coast (except in hot tropical lands where the sea breeze can penetrate further and faster). In the United Kingdom and much of Europe the majority of sea breeze soaring flights have been within 25 miles of the coast.

Dependence on offshore flow

The front is unlikely to give much lift unless the early morning flow has a small offshore component. It is the opposition between land- and sea-air flows that produces the convergence line along which lift develops.

Appearance: curtain clouds

See fig. 116(a). The front is often marked by a line of cumulus, sometimes much larger than the clouds further inland. The distinguishing feature is a ragged line of 'curtain cloud', rather wispy fragments which seem to hang down below the main base. These wisps form where the cooler and moister sea air starts to

rise at the front. The best lift normally occurs very close to the curtain clouds, just on the landward side.

Nearly all the cloud disperses on the seaward side of the front, though when it penetrates far inland feeble cumulus with a lower base may develop. These clouds mark small weak thermals which seldom last long. The lift is apt to be weak and broken.

If the air is too dry for cloud the line of the sea-breeze front may show up as a change of visibility. Figure 116(b) shows what the sea air might look like if there was a great deal of murk in it. The haze is very rarely dense enough to reveal all the bulges and undulations at the sea-breeze front.

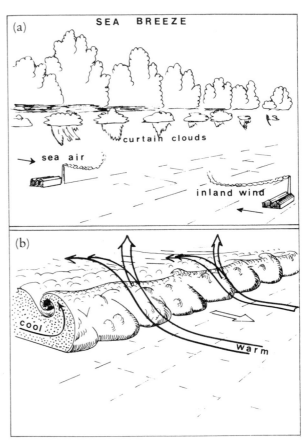

116. (a) and (b) Sketches of sea breeze

What the air feels like

The front usually produces a narrow line of strong lift with some turbulence associated with the curtain clouds. When clouds and visibility changes are lacking, flight through the front produces a different feel to the air. A motor glider pilot flying at 2,000 feet

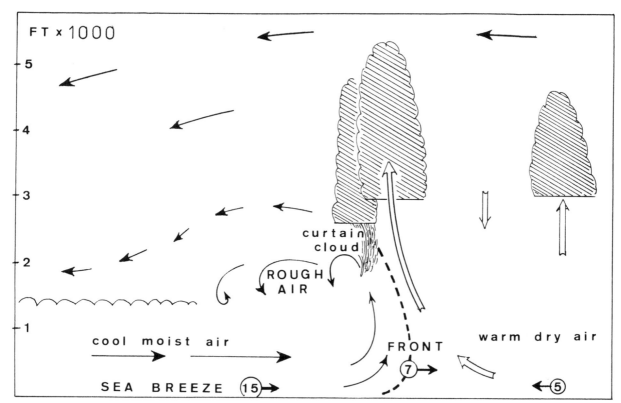

117. Cross-section of sea breeze

through a sea-breeze front described conditions as 'rough' near the front, then 'bubbly' and finally 'smooth'. On the seaward side of the front thermals are almost completely absent and the air feels dead. Figure 117 illustrates a sea-breeze front in cross-section.

Sea-breeze fronts and showers

Nearly all sea-breeze fronts increase the size of cumulus because the normal strength of thermals is boosted by the line of extra lift along the front. The effect is most marked when there is enough of an offshore wind to slow down the incoming front and increase the upward flow. On some days the extra lift has boosted Cu tops many thousand feet above the level predicted from nearby upper-air soundings. With an irregular coastline sea-breeze fronts may move inland from two or three directions. Where they meet the convection can be particularly strong; in a few cases when the air was already very unstable thunderstorms have broken out at the intersection points.

Pseudo sea-breeze fronts

The supply of cold, moist air behind a 'sea-breeze front' does not always come off the sea. A similar line of convergence develops on calm days when one part of the country lies under a thick sheet of cloud or fog while bright sunshine warms an adjacent region. During the afternoon the contrast between the sunny region with cumulus clouds and the cold grey overcast area produces a circulation just like a sea breeze. A wind starts to blow from the overcast to the sunny region and a convergence line, similar in nearly every respect to a sea-breeze front, develops at the edge of the sunny region. Such 'pseudo sea-breeze' fronts move from cold to warm areas and may produce a sudden change in the weather when they pass.

In one particularly bad case the front carried thick overnight fog some thirty miles across a previously clear region during the middle of a sunny afternoon. Nobody had expected fog to arrive at such a time and a training airfield had to divert a number of pupil pilots from their home circuit to an airfield still in the clear.

How the shape of the coast affects the sea breeze

Figure 118 shows a series of diagrams to illustrate the various orographic effects.

(a) shows a straight coastline in the morning when the land is colder than the sea and an offshore wind blows.

(b) shows the change by mid-afternoon when the land is much warmer. The sea breeze has penetrated inland and a front has formed where it meets the wind off the land. Out to sea there is a region of sinking air where the flow diverges, some curving inland and some heading away from the coast.

(c) shows how sea air penetrating inland from either side of a peninsular can produce a convergence line in the middle.

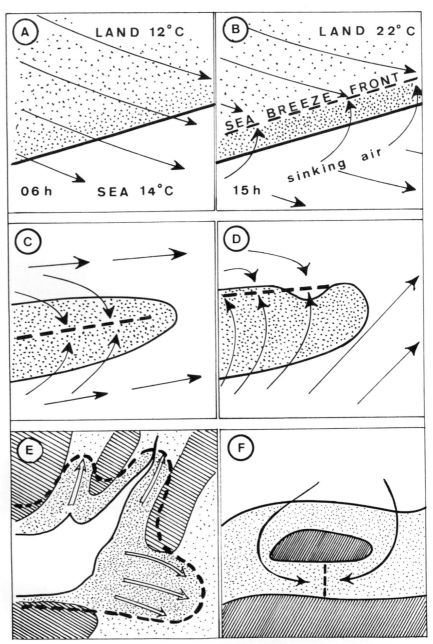

118. *(a) to (f) Sea breeze locations and penetration inland*

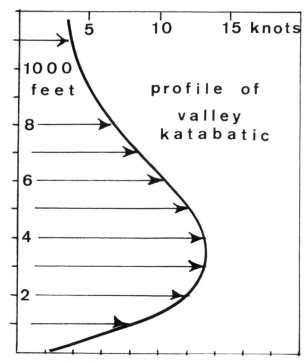

119. *Profile of katabatic wind flow down a valley*

and snow, katabatic winds can reach gale force. However, in most places the wind speed is usually less than 15 knots.

Figure 119 shows the vertical profile of a katabatic wind flowing down a valley. The wind only affects a shallow depth of air. In this example the wind was strongest at 300–400 feet and decreased to less than 5 knots above 1,000 feet. The anabatic upslope wind is usually very much shallower.

Valley winds

Mountains form a strong barrier to the winds at low level. The air tends to flow up or down the valley rather than over the intervening ridges. The effect is most marked when the valley air is cold and stable. Figure 120 shows a series of sketches to illustrate winds in a mountain valley.

(d) is a variation of (c) which occurs when the wind originally blows across the peninsular. The sea-breeze front is then found near the lee coastline and may sometimes be found out to sea, especially where there is a small bay.

(e) illustrates the penetration of sea air up a flat estuary between areas of high ground (which are shaded in). The sea air makes good progress over the low ground and penetrates up valleys or through gaps in the hills. The speed over high ground is much slower.

(f) shows how the sea breeze may encircle an isolated mass of hills to produce a convergence-zone well inland on the far side of the hills.

Anabatic and katabatic winds

When long slopes are heated by the sun the air above them is also warmed and a wind starts to blow uphill; this is called an anabatic wind. The opposite effect occurs (particularly during the hours of darkness) when the slopes lose heat by radiation. The chilly air descends the slopes to produce a katabatic wind. In some arctic and antarctic regions, where there are large masses of high ground covered with ice

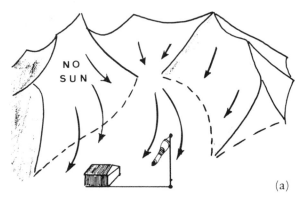

(a)

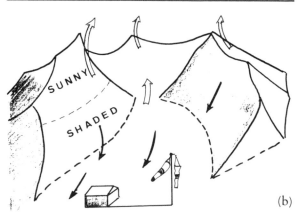

(b)

120. *(a) to (e) Development of mountain and valley winds throughout the day*

(a) shows conditions at dawn when overnight cooling has produced a katabatic flow down the mountain slopes into the main valley and thence down the valley and out towards the plains.

(b) After sunrise the upper slopes start to warm and the heated air begins to rise up them. The valley and lower slopes are still in shadow so the katabatic wind continues at low levels but starts to weaken.

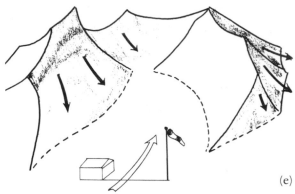

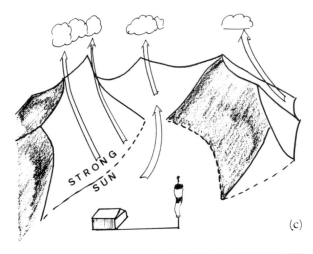

(c)

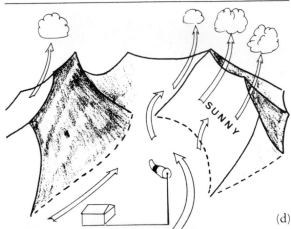

(d)

(e)

descent in the high valleys and the end of the anabatic from the plains, but eventually the flow pattern reverts to the nocturnal pattern shown in (a).

Figure 121 shows a section of mountains and valleys to illustrate the cross-valley flow through the day:

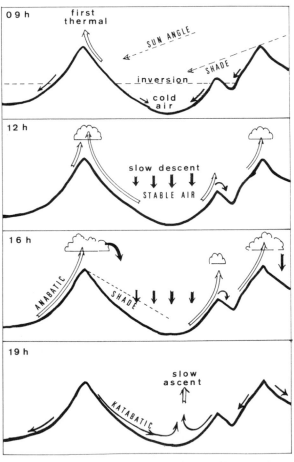

(c) Just before midday, when most of the slopes are in sunshine, there is widespread ascent of air from the peaks and ridges and an anabatic flow up the valley to replace the ascending air.

(d) During the afternoon the location of the main upcurrents changes as the sunshine warms different slopes. The valley wind increases in strength and draws in an extra supply of air from the plains.

(e) At sunset the mountain slopes cool off rapidly and the air starts to sink down the slopes towards the valley. There is often a lag between the beginning of

121. *Cross-section of mountain and valley winds 0900 to 1900 hrs*

At 0900 hours the valley is still filled with cold air topped by a temperature inversion which traps any haze or fog. The sun has just begun to produce anabatic winds on the sunny upper slopes with thermals leaving the ridge line.

At 1200 hours there are well-developed anabatics up the mountain sides and a slow descent of air over the valley keeping conditions stable there.

At 1500 hours east-facing slopes are in shadow and the anabatic flow is chiefly confined to the west-facing slopes.

At 1900 hours katabatics have begun to develop down all the slopes. Where they meet over the valley there is sometimes a zone of slowly ascending air. This evening lift has been explored by sailplane pilots; it is usually too weak to climb in but can be used to extend the final glide at the end of the day.

13

Airflow over ridges and mountains

Air flows round and over mountains in many different ways. Most weather maps are drawn on far too small a scale to show what is actually observed by hang gliders, sailplanes and any other aircraft which may fly close to the ground. This section illustrates some of the features which may be encountered.

Flow across ridges

Figure 122 shows simple flow patterns over low ridges.

(a) is a smooth windward slope with no sharp edges. The dotted lines outline the region where lift is usually found.

(b) is a steeper ridge with a sharp edge at the top where the flow separates. Separation produces an eddy over the flat-topped hill; in such conditions landing and take-off from a hill-top airfield can be difficult. The eddy produces an effect termed 'curl-over' which often causes much turbulence at low levels. It may temporarily reverse the wind direction on the airfield. Sailplanes have been rolled through

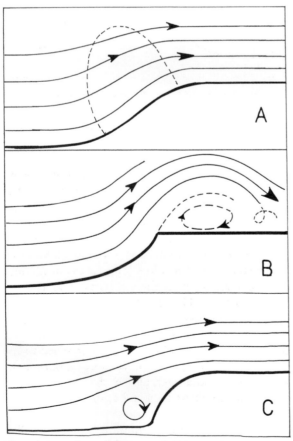
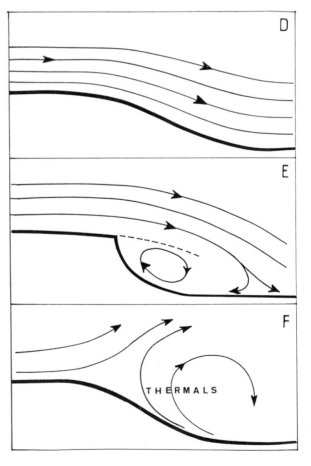

122. (a) to (f) Airflow over different ridges

nearly 90 degrees on descent through the curl-over. This has caused accidents when it has happened near the ground.

(c) illustrates a ridge which rises very steeply from the base. This often traps a local circulation called a 'bolster eddy' at the foot of the slope. The rotation in this eddy makes the air actually descend the face of the slope at low levels although higher up there is normal upslope motion. Pilots who approached the ridge at low levels expecting to be carried upwards have run into a region of sink which forced an unintended landing at the foot of the ridge.

(d) is a smooth lee slope down which the air descends without turbulence. Extra height and speed are needed when flying across this zone.

(e) is a sharp-edged slope where the airflow separates at the top leaving a lee eddy underneath. This lee eddy produces a narrow band of rising air tucked in very close to the lee slope. Seagulls may be seen to take advantage of this eddy to rise almost to the top of the lee cliff. Anything much larger would have problems using this lift.

(f) shows that usable lift may be found near a lee slope if the sun has warmed it enough to set off thermals. These thermals break up the normal region of sink on the lee side. They are sometimes called 'wind shadow thermals' and they may form because the air on the lee side has been left undisturbed long enough to become extra warm but it is possible that lee wave activity has triggered off the thermal.

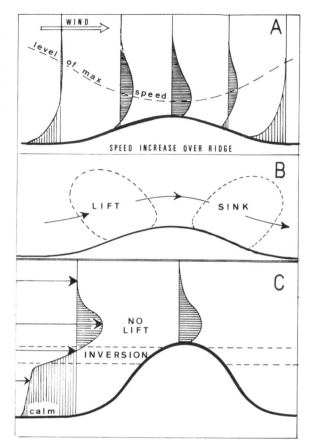

123. *(a) and (b) Increase in wind speed over a smooth ridge*
(c) Cold air below an inversion blocked by a ridge

How the wind speed alters with height

Surface drag reduces the wind speed near the ground. Over flat ground the vertical profile of the wind speed has a marked curve at the bottom where the effect of drag is most marked. When the air flows over a smooth ridge the wind speed often increases. This is illustrated in fig. 123. In this set of diagrams the wind blows from left to right and the speed profile is shown by a series of upright lines. If the wind speed is constant with height the profile is just a straight line. Where the wind speed has been increased there is horizontal shading. Where the wind speed is reduced the shading is vertical.

(a) shows that the level of maximum wind dips down where the air passes over the ridge. The horizontal shading shows a marked increase of speed just above the ridge. The effect is similar to the flow over the top surface of a rather thick aerofoil. Upwind of the ridge the profile shows how surface drag has slowed the air down near the surface; the same effect is seen developing again just beyond the crest of the ridge and again over the flat ground well to leeward.

(b) shows the regions of lift and sink to be expected. Notice that at low level the usable lift does not begin the moment the air starts to ascend the windward slope. The drag has so reduced the speed near the ground that the lift is too small at the foot of the ridge. It improves as the wind speed increases higher up. Since this effect is chiefly confined to the layer nearest the ground it is more important to model aircraft and hang-gliders than to sailplanes. Most sailplane pilots tend to keep further out from the slope.

(c) shows a different situation; here the valley is filled with cold air capped by a temperature inversion. The effect of an inversion is to isolate the cold air beneath it from the warmer air above. The wind speed is usually very much less below the inversion and may be calm at the surface. Just above the inversion the speed can be stronger than average (it is as if the stronger wind above the inversion was compensating for the very light wind below). If the inversion is up at ridge-top level the colder air in the valley does not flow up the ridge; it is said to be 'blocked'. Thus, although a pilot standing on the top of the ridge can measure a breeze which should be enough to give slope lift it does not actually become soarable until the air in the valley has warmed up enough to break down the inversion. On days like this observers at hill-top airfields may find the wind speed decreases between dawn and mid-morning; at nearby flat sites the wind is lightest at dawn and strengthens during the day.

Isolated hills

The flow over isolated hills is not as simple as over a long ridge; much of the air can go round the side instead of over the top. Figure 124 is an example of a conical hill rising out of the sea. There are no obstructions upwind so the approaching flow is smooth.

(a) shows a plan view of the peak with a wide arc showing the base and a smaller ring showing one of the upper contours. The approaching wind direction is given by the three wide arrows blowing from the WSW. A series of pecked lines show how the flow curved round both sides of the hill to meet on the lee side. There many of the streamlines turned back to rise up the lee side.

(b) is the side elevation of the hill. The streamlines dip as they follow the sides of the hill and then curve back under the lee eddy and rise towards the point of convergence shown on the plan. From this point the flow breaks away in a twisting path marked 'vortex trail'. Trails like this have also been observed in water tank experiments. The region of rising air on the lee side is only likely with fresh to strong winds and the pattern does not remain steady. It is liable to break up and then reform in a slightly different place.

Figure 125 shows a rather different pattern over a big conical mountain.

(a) is a cross-section of the mountain. Once again there is a lee eddy whose effect is to produce a region

of lift extending some distance downwind of the peak. In moist conditions a banner cloud trails away downwind of the peak.

(b) gives the plan view of the peak. The main regions of lift were found in three lobes spaced at equal intervals round the peak.

(c) is a map of Tenerife with arrows marking the flow round the sides and up to form a banner cloud. Banner clouds have been seen blowing from a number of peaks; Everest and the Matterhorn are two of the better-known examples.

Air flow deflected by a ridge

The foehn wind was described in Chapter Eight. (Figure 126 is repeated to save turning back to the earlier chapter.) The upper diagram shows the typical foehn wind ascending and giving up most of its moisture over the windward slopes and the crest of

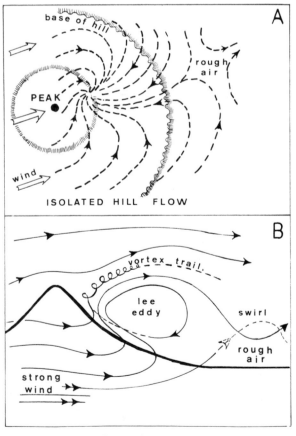

124. *(a) and (b) Plan and cross-section of flow over and round an isolated hill*

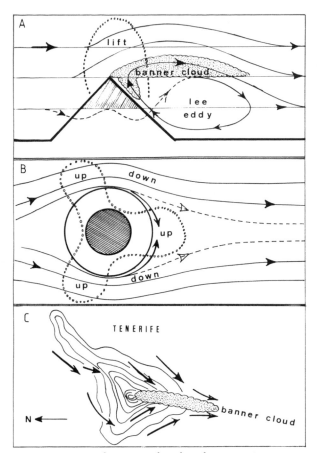

125. (a) to (c) Flow round isolated mountain producing banner cloud

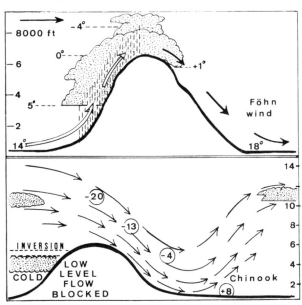

126. Foehn winds: ascent and descent with drying, low-level flow blocked

the ridge and then descending the lee side as a much drier current. The lower section shows the low-level flow blocked so that only the high-level flow descends the lee side.

Figure 127 is a 3-D drawing to represent what may occur when the low-level flow is blocked by the ridge. When this air is too cold and stable to climb over the ridge it is often deflected so as to blow along the ridge towards the low-pressure region. The warm, less stable air passes over the cold flow at right-angles as it goes across the ridge line.

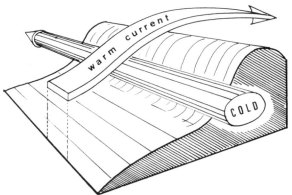

127. Blocked low-level wind deflected parallel to ridge

Anabatic winds and mountain thermals

Figure 128 illustrates the temperature changes as air ascends the sunny slopes of a mountain. The height in thousands of feet is marked up the left side of the diagram together with the temperature of the surrounding air. For this example the environment temperature is assumed to decrease at 2°C per thousand feet. A thermal rising from level ground would cool at 3°C per thousand feet and would not rise far in this environment. If, instead of rising vertically, the thermal keeps close to the mountain side it will gather heat continually from the sunny slopes. The temperature of the air above the slope is shown up the mountain side. On the right is a column marked 'DIFF' which is the temperature difference between the mountain thermal and the environmental air.

As the thermal rises the temperature difference (and hence the density difference) increases and at the 6,000 feet level it is 3 degrees. Then the thermal

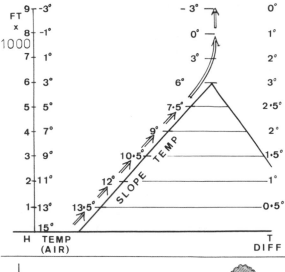

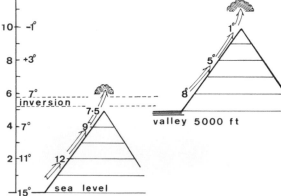

128. Anabatic wind; slope thermal

The form of mountain thermals

Figure 129 shows a more detailed picture of mountain thermals.

(a) shows how the air rises up the sunny side of the mountain when there is a wind blowing from left to right. Wind speed increases gradually with height. The thermal sticks close to the slopes until there is a small dip. There the thermal breaks away but is later joined by another slope thermal starting higher up. Together they form cumulus above the peak. The cloud is constantly being reformed as earlier bits drift away downwind and decay.

(b) illustrates the effect of a greater increase in wind speed aloft. The first breakaway thermal decays while the second (which started higher up) forms a series of detached cumulus to lee of the peak: these are rapidly blown away downwind. Instead of find-

breaks away from the peak and ascends vertically. Now it cools at the standard 3°C per thousand feet and at the 9,000 feet level it is no longer warmer than its environment. This shows how the anabatic wind can produce lift up a sunny mountain slope even when the surrounding air is stable.

(b) compares the difference between a coastal mountain and one rising from a valley floor of 5,000 feet. On this occasion there is an inversion just below 6,000 feet. The coastal mountain drawing up air with a higher dew point produces a cumulus above the summit at about 6,000 feet. The inland mountain starts with the advantage of slightly warmer and drier air at valley level. When this ascends the slopes cloud does not form until the air reaches 11,000 feet.

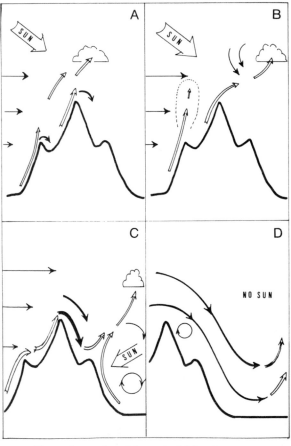

129. Slope thermals with a wind blowing across the mountains

ing continuous lift from peak to cloud base the lift is interrupted by patches of sink.

(c) Here the sunshine is on the lee side of the mountain. On the windward side there is some upslope lift (provided the slope is not too steep) but at the summit the air spills over to produce heavy sink on the lee side. Lower down the mountain side thermals provide lift up to a mid-way hump where they break away. The region of lift then tilts downwind to form a cumulus well away from the main peak.

(d) Once the sunshine has ceased air tends to descend all the way down the lee slope, probably with eddies below the steeper sections and perhaps producing wave flow over the far side of the valley.

The depth of slope thermals: how the lift varies

The anabatic wind which develops when thermals rise up the face of a slope is usually much shallower than the katabatic which flows down the valley at night. Figure 130(a) is a cross-section of the anabatic wind. Most of the strength is within 100 feet of the slope and the best lift is only about 50 feet away. This is why experienced pilots tend to fly very close to the mountain side.

The best lift is usually found over a smooth slope without too steep a gradient. (b) shows weak lift beside an almost vertical cliff with strong lift close by where the angle is less.

A large cloud shadow persisting over one section of the ridge may interrupt the anabatic wind (c). Sometimes thermals break away at the lower side of the shaded region. (d) illustrates the thermal breaking away from the face over an intermediate peak leaving a curl over the region of sink beyond. When the mountain rises in steps like this it is wise to gain extra height before crossing these sections.

Effect of a subsidence inversion

Figure 131 illustrates the effect of an inversion which develops well above the mountain tops initially but

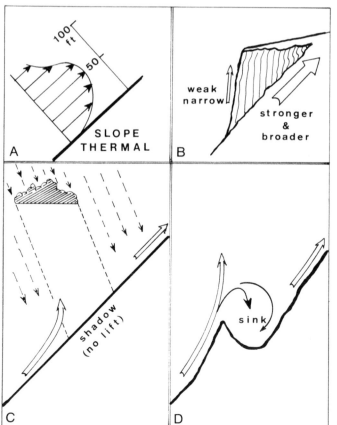

130. *(a) to (d) Details of an upslope wind*

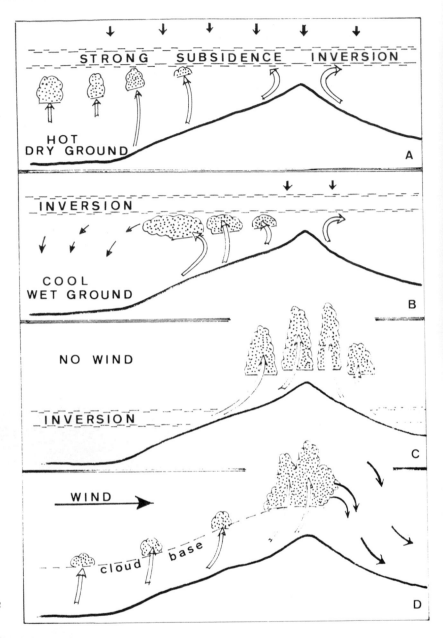

131. Influence of inversion on mountain thermals

then settles down lower day by day. This sequence may occur when a ridge or anticyclone moves in after the passage of a front. The first three diagrams are for nearly calm winds.

(a) shows a strong subsidence inversion above the mountain peaks. If the valley is very dry there may be thermals rising over a wide area. The cloud base will rise over the high ground but the inversion will restrict the tops. On such a day Cu become smaller and flatter as you ascend the slope until they disappear completely near the summits.

(b) is the same situation except that the low ground is too moist to give thermals. Lift is then restricted to the higher slopes and a circulation develops with air flowing outwards under the inversion and sinking down into the damp valleys.

(c) Now the inversion has moved lower so that the upper half of the mountains are above it. Thermals can then develop strongly over the peaks but the valley air (beneath the inversion) remains lifeless.

(d) The inversion has been removed and a wind is blowing in from the left. The base of cumulus rises as you travel from the plains towards the higher mountains. Thermals are assisted where the wind blows up the slopes. As there is no restraining inversion the cumulus can grow much larger over the higher mountains. On the lee side cloud decays rapidly and a wide region of sinking air prevents thermals from forming.

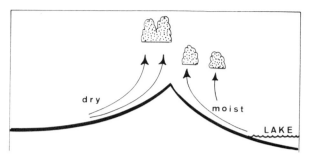

132. *Differences of cloud base on either side of a ridge*

Differences in cloud base

Mountain ranges often separate air masses with very different humidity. On the coastal side of a major range cloud base is usually much lower, especially when moist sea air is drawn in during the day. On the other side where the air is drier the cloud base will be much higher. The effect is also found when there is a large lake on one side of the range, as shown in fig. 132.

The difference can be so great that an entirely different climate develops on the landward side of a coastal range. This makes cross-country flights difficult to plan in mountainous regions. The pilot may start in the dry region with strong sunshine and a cloud base well above most peaks, then at the divide there is a complete change and the pilot looks down on the tops of an almost unbroken cover of cloud. This is a common experience in Alpine regions but may be encountered with the far smaller hills of Scotland and Wales.

Mountain cumulus over the Austrian Alps, south of Zell am See. The cloud was forming over the high ridges; over the valleys there was only sinking air

14

Visibility

How it is defined and measured

The visibility reported by Met. stations is defined as the greatest distance at which a dark object can be seen against the horizon sky. This is fairly simple to determine by day but rather difficult by night. One way to get consistent results is to use a photo-electric cell to measure the decrease in intensity of a beam of light travelling a known distance. This gives a standard for comparing visibility by day and night.

Runway visual range

When the visibility is poor (1,500 metres or less) it is more useful to report how far specified runway lights or markers can be seen by a pilot about to touch down. This is measured from a point close to the runway at major airports and reported as 'runway visual range' (RVR). When there is mist or fog at night the RVR may be better than the official visibility because a light can be seen further than an unlit obstruction.

Reasons for reduced visibility

(a) Dust particles causing 'haze'.
(b) Droplets of water causing mist or fog.
(c) Small particles of sand lifted by strong or gusty winds.
(d) Precipitation of any sort: in ascending order of severity this may be rain, hail, drizzle and snow.
(e) Smoke from manufacturing regions, domestic fires and stubble burning.
(f) Sea spray (chiefly over the ocean but also carried inland by gales). Spray does not usually cause serious visibility problems unless the wind speed rises to storm force.

The importance of inversions

A temperature inversion, where relatively cold air is capped by a layer of warmer air, often acts as a lid to prevent interchange of air at different levels. Most of the dust, smoke and chemical pollution is produced at or very near ground level. If the air is unstable, convection currents rising from the warm ground carry the haze upwards and spread it over a deep layer of the atmosphere. The deeper this layer becomes the less is the concentration of haze. However, when an inversion stops convection or limits it to a very shallow layer the haze becomes concentrated underneath and makes the visibility much worse.

Haze

Haze is produced by particles of dust, smoke, chemical pollution and even grains of pollen carried up by convection currents. Much of the haze over Europe and the British Isles is due to industrial pollution trapped beneath an inversion. The worst conditions are found downwind of a major industrial region such as the Ruhr in Germany or the heavily industrialised areas of Czechoslovakia. The wind carries this haze for hundreds of miles. Visibility is apt to be much worse in winter than in summer because more smoke is formed then and the sun's heat is seldom strong enough to lift and break the inversion. During the summer months daytime heating often lifts the inversion and thermals spread the haze over a deeper layer. The visibility is best during the period of maximum heating and deteriorates at night when radiational cooling produces a new inversion just above the surface. Figure 133(a) shows how the haze is lifted and thinned out during the day and also how the visibility may get worse higher up just under the inversion.

Haze is worse when the air is humid

Some haze particles are hygroscopic, that is they absorb moisture even when the air is not completely saturated. This increases the size of the particles and makes the visibility worse. An increase of humidity often goes with a decrease in visibility. The effect is most noticeable when cumuli form below the inversion. For two or three hundred feet below cloud base the humidity is high and on hazy days the visibility deteriorates at that level. This can increase the risk of collision between aircraft flying just under the cloud base.

Cloud tops and haze tops

(a) **Small cumulus** The inversion which traps the haze also limits the growth of small cumulus. Clouds which form at the tops of strong thermals often push through the inversion for a few hundred feet before they lose momentum. As a result the higher tops rise into clear air. A glider can use such clouds to rise (briefly) above the haze layer and catch a glimpse of distant clouds. After several hours of thermals the top of the haze may rise over a thousand feet.

(b) **Stratocumulus** Large areas of Sc are often found above winter anticyclones. The air above the cloud is stable. Heat radiates into space from the top of the Sc layer which becomes progressively colder. This cooling increases the temperature difference and produces an inversion which grows stronger day by day. Haze fills the layer between cloud and ground but the visibility does not change much between day and night. This is because the cloud acts as a blanket preventing warmth from the ground radiating away into space. The temperature stays fairly constant by

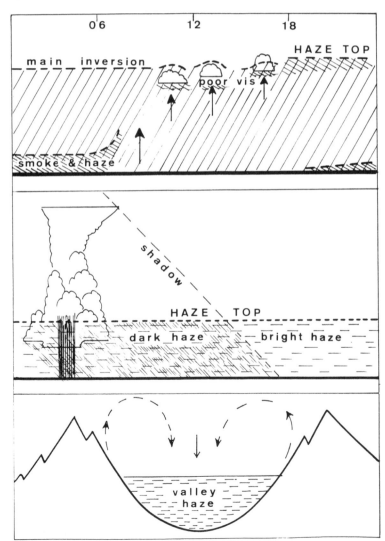

133. *(a) Haze variations during the day*
(b) Cb shadows on haze layer
(c) Haze in valleys

day and night so the visibility does not get significantly worse overnight.

(c) **Big cumulus** Towering cumulus clouds usually build through the haze layer and rise many thousand feet above it. The haze top is not usually carried up by these clouds. The clouds cast shadows on the haze making it look dark. This darkening of the haze on a hot summer afternoon is sometimes the only warning of the proximity of a developing thunderstorm. Figure 33(b) illustrates this.

(d) Dust is sometimes carried up in **layered frontal cloud** to 20,000 feet or more. When this cloud evaporates a thin dust layer remains. Such layers are usually so thin that they are not noticed from above or below and are only seen as a brownish layer when you are level with them.

Valley haze

Haze is often trapped in deep valleys, especially during the winter months when little sunshine reaches the valley floor. There is an inversion separating the cold, haze-filled valley air from the clear air aloft. During the day thermals rise from the haze-free peaks but this air descends over the valley to maintain the inversion.

Dust and sand storms

In dry, desert regions the squall line associated with an approaching cold front can raise a wall of dust and sand and carry it hundreds of miles. Vigorous depressions which form over North Africa and the Middle East have been known to lift desert dust to heights of 30,000 feet or more where it mixes with the upper clouds. Southerly winds sometimes carry this dust across Europe to the British Isles and even Scandinavia. It is eventually brought down as 'coloured rain'.

The low-level outflow from thunderstorms produces a less widespread duststorm known in the Sudan as a 'Haboob'. Similar duststorms occur in many desert regions.

Smog

The word is a combination of smoke and fog. Industrial pollution combined with fog makes the visibility particularly bad. In Britain the Clean Air Act has reduced industrial pollution so much that the visibility is often better in London (where temperatures are higher) than out in the fog-bound countryside. The term is also used for a photochemical haze which develops by the action of sunlight on car exhaust fumes and factory emissions. Smog is trapped in cold air below inversions and becomes progressively thicker day by day if the air becomes stagnant in a large hollow between mountains. Parts of California are notorious for smog but in recent years Mexico City has gained a reputation for severely polluted smog.

Differences between ground and air visibility

In hazy conditions one can see much less looking into sun than looking down sun. ATC may report the visibility as 5,000 metres but when flying into sun an airfield may be invisible until you are almost overhead; looking down sun one can pick out runways and hangars 3 miles away.

The effective visibility often changes during a let down to circuit height. Haze which is lifted up during sunny days can become thicker beneath the inversion. ATC may report '10 kilometres or more' but the pilot is unable to see the airfield at that distance until he has descended well below the inversion. Then everything becomes clearer.

The reverse effect occurs when there is a very shallow inversion near the ground. The air is often very clear at flying levels and the airfield is easy to see until the last few hundred feet before touchdown. In extreme cases a pilot has let down into a very shallow layer of thick fog which had not been seen from aloft. This is fortunately rare but it can occur in very cold, calm conditions.

Air-to-air visibility

In the absence of cloud the air-to-air visibility almost always improves with height. The change in visibility is most marked as you climb above an inversion. The layer of smoke and haze sometimes looks thick enough to walk on. The ultimate inversion is the tropopause, the dividing line between the lower atmosphere (troposphere) and the stratosphere. The tropopause may be seen during high-altitude flights as a distinct change in clarity.

Mist and fog

When air is cooled below its dew point the moisture condenses out as minute droplets of water. (In cold countries some fog may be due to ice crystals.) The water droplets form cloud in the free atmosphere, dew on cold surfaces and fog in the layer of air touching the ground. Airfield reports use the term mist when the visibility is more than 1,000 metres and fog when it is less. In the USA the term 'slight fog' is used instead of 'mist'. News bulletins directed at the motoring public generally confine the term fog to visibilities of 200 metres or less.

Basic causes of fog

Fog forms when the air remains in contact with a cold surface. The surface must cool the air below its dew point but this alone is not enough. Dew forms on many clear nights but fog does not always follow. There are several classes of fog:

(a) **Radiation fog** usually occurs under clear skies when the ground loses heat by radiation. The air is cooled by contact with the ground and fog develops if there is sufficient moisture in the lowest few hundred feet. This fog does not usually form unless the wind speed is very low. On many occasions it is patchy, being confined to low ground where the coldest air settles during the night. The extra moisture in valleys also helps fog to form.

(b) **Advection fog** occurs when the wind blows relatively warm moist air over a cold surface. Unlike radiation fog it can form in strong winds and travel long distances. It is most likely when the air from sub-tropical oceanic regions is brought polewards over progressively colder seas. It also occurs in the winter months when warm air travels over snow fields. This sets off a slow thaw and widespread fog develops.

Sea fog is a form of advection fog; it is more common during spring and early summer when the sea is still relatively cold. Satellite pictures show that sea fog often has a very clear-cut edge. The boundary is rarely straight but has indentations where eddies of dry air have been drawn into the edge of the moist flow. Sea fog is carried inland during the day when sea breezes develop. It may arrive as very low cloud at airfields near the coast. Sea fog can be very persistent over regions where the sea temperatures are particu-larly low. There are patches of cold water along the east coasts of England and Scotland where a narrow zone of sea fog persists long after the rest of the North Sea has cleared.

(c) **Hill fog** may be due to low cloud forming on hills or because an area of radiation fog is advected up gentle slopes by a slowly increasing wind. This is also termed 'upslope fog'; it is a hazard to early morning flying at high-level airfields because it often arrives an hour or two after sunrise on a previously clear morning when the wind starts to stir up overnight fog which had been confined to low ground.

(d) **Steam fog**, so called because it looks as if a water surface is turning into 'steam', occurs when cold air flows over water which is much warmer. The lowest layer of air is quickly warmed by contact with the water and picks up extra moisture. The warming sets off rising convection currents. As soon as this newly-warmed and moistened air rises it mixes with the much colder air above and the extra water vapour immediately condenses out. The process works best when the air arrives with a temperature well below freezing; since it can hold very little water vapour at such low temperatures it quickly becomes saturated.

Steam fog is sometimes seen over streams and rivers on calm cloudless evenings but is much more common in arctic regions where it is termed 'arctic smoke', 'sea smoke', or 'frost smoke'.

The development of radiation fog

The processes involved in radiation fog are not entirely simple; there is often a delicate balance between opposing features. As a result fog is often patchy and forecasts of fog formation are liable to considerable timing errors. It is difficult for a Met. Office responsible for many different airfields to make allowance for all the variations of terrain and the small differences in temperature and humidity. Landing forecasts express this uncertainty by quoting a broad range of times for fog formation or adding a percentage probability of fog with phrases such as 'Prob 30 tempo...' (30% Probability of temporary...).

Stages of fog

Figure 134 illustrates the formation and thickening of radiation fog. Isotherms show the temperature in the lowest 800 feet from mid-afternoon through to the following morning. The depth of fog is shown by shading. The stages in fog formation are as follows.

1. When sunlight fades the ground cools by radiation out into space; this is greatest when there are no clouds.

2. Air in contact with the ground cools and a temperature inversion is formed. On some calm, clear nights the temperature measured on the ground may be as much as 5°C below the temperature in the thermometer screen 4 feet higher.

3. As the inversion strengthens it decouples the wind aloft from the wind near the surface. There is no longer a smooth decrease of wind speed with height; instead the speed decreases sharply beneath the inversion.

4. A lull in the wind (a drop from about 4 knots to less than 1 knot) stops all turbulence, and surface cooling becomes more rapid. If the wind should increase again the cooling slows down.

5. The cooled air becomes saturated near the surface; as cooling continues the excess moisture is deposited on the surface as dew. This removes some of the water vapour from the air. For dew to be deposited there must be slight turbulence to mix up the air and maintain a supply of moisture at ground level. Once the wind drops to about 1 knot the absence of turbulence checks the deposit of dew.

6. When moisture is no longer deposited as dew further cooling results in a very thin layer of fog forming just above the ground.

7. The fog becomes deeper with time and then acts as a blanket shielding the ground from radiation cooling. Loss of heat now takes place from the top of the fog. The fog top rises as it cools.

8. Meanwhile, warmth in the upper layers of the soil is conducted up to the cold surface and checks the fall of temperature there. With the fog top cooling but the ground temperature steady, the layer of fog becomes slightly unstable. Very weak convective currents then develop, stirring up the fog very slowly and producing humps and valleys in the fog top.

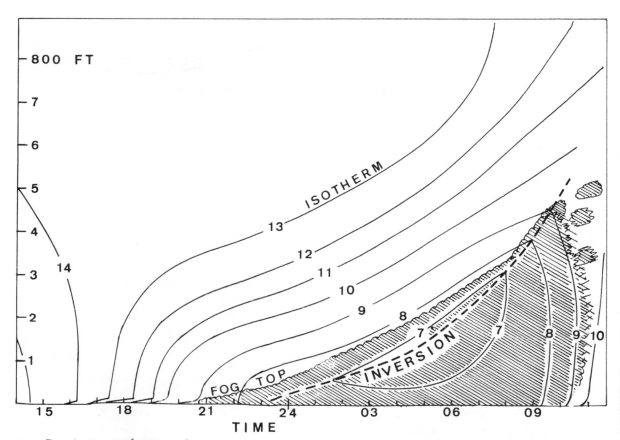

134. Development of autumn fog: a time cross-section showing increase of depth throughout the night

9. On a deep layer of fog gravity waves may move along the inversion producing regular undulations.

10. If the air is motionless the fog may not start to deepen until the first stirring of wind around sunrise. On some nights the fog remains very shallow for hours and only thickens at sunrise. Then the fog deepens rapidly and may spread out over a larger area.

Outside influences

While these delicate adjustments are taking place the region of fog may be disturbed by other factors.

(a) A strengthening katabatic wind may move the fog away as drier air flows down from high ground.

(b) The arrival of a cloud layer may stop fog forming or clear it away after it has developed.

(c) Freshening winds aloft may start to move the fog bodily towards previously fog-free areas. Sometimes this movement is up a very gentle slope towards higher ground. The air cools slightly during the ascent, helping to maintain the layer of fog. As a result high-level stations may be blanketed by fog which originated over flat, low ground. On some mornings this effect clears the overnight fog from low ground by shifting it to the higher ground where it persist for several more hours.

Figure 135 illustrates these effects.

(a) is simple valley fog formed where the cold air has drained off the hills.

(b) shows how a katabatic wind may move the fog away from the foot of the hills.

(c) illustrates how the fog may be displaced up the sloping side of a hill.

(d) is true upslope fog where the air is cooled by the slow ascent.

Radiation fog and the winds aloft

It needs to be almost calm at the surface before radiation fog will form. There is always a decrease of wind between 2,000 feet and the surface but the turbulence associated with strong winds tends to keep the air too well-stirred for a surface inversion to develop. Until this inversion forms the surface wind cannot become light enough for radiation fog. Over England the critical speed for radiation fog seems to be 17 knots at a height of 2,000–3,000 feet and about 5 knots at 33 feet.

If the upper wind is less than 17 knots a nocturnal inversion can form and this allows the 'surface wind' (which is actually measured about 30 feet above the airfield) to drop below 5 knots.

Dispersal by strengthening winds aloft

Once the air near the surface has been isolated by the inversion the upper winds can increase considerably before they disperse the fog. On some days the 2,000-foot wind increased to just over 30 knots before the fog cleared. Table 6 shows the 2,000-foot wind speed divided into 3-knot steps with the percentage frequency of nocturnal fogs in each step. Column (a) shows the wind before fog formed, column (b) the wind speed reached before fog cleared.

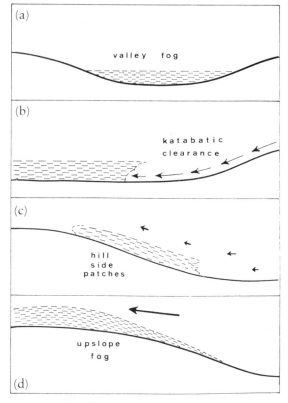

135. (a) Valley fog
(b) Katabatic wind clearing valley fog
(c) Hill side fog
(d) Upslope fog

Table 6. Fog and the 2,000-foot wind speed: (a) when fog first formed; (b) before fog cleared

Speed range (knots)	Percentage of fogs	
	(a) when formed	(b) before dispersal
00–02	7.1	6.7
03–05	16.7	16.0
06–08	23.8	18.0
09–11	23.0	20.0
12–14	18.3	16.3
15–17	10.3	11.3
18–20	0.8	6.7
21–23	—	2.5
24–26	—	1.5
27–30	—	0.5
over 30	—	0.7

Fog and surface wind speeds

There can be a significant change in wind speed between the height of 2 metres, where fog first forms, and the 10-metre wind which is reported by ATC. Radiation fog is unlikely to form if the 10-metre wind is more than 5 knots. (The speed at 2 metres will probably be only 1–2 knots.) The exceptions are when there is both cooling by radiation and advection of moister air. For example, in central and eastern parts of England a light wind off the North Sea brings extra moisture and fog may persist as a mixture of radiation and advection fog with 10-metre wind speeds as much as 13 knots.

Table 7. Wind speed at 10 metres (33 feet) during fog in central England

Speed range (knots)	Percentage of fogs
00–02	58.5
03–05	28.6
06–08	10.5
09–11	2.1*
12–14	0.3*
15–17	—

Asterisks show occasions when radiation fog formed but wind then increased as sea fog was advected inland. There was no break between radiation fog and sea fog.

Weather patterns for fog

Anticyclones, ridges and cols are the regions where radiation fog is most likely. These are all regions

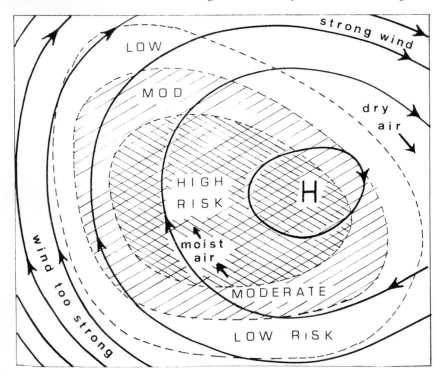

136. *Radiation fog and anticyclones; degrees of risk related to position of the high centre*

where there is a fair prospect of clear skies at night and the isobars are spaced far apart so that the 2,000-foot winds are light.

The risk of fog is greater when the winds bring increasingly moist air. In the northern hemisphere light winds from a southerly direction are more of a fog threat than northerly winds. This means that the western side of an anticyclone is often more foggy than the eastern side. Figure 136 shows how the risk of fog varies in different sectors of an anticyclone. The pattern is not symmetrical with the isobars. The western side of the high is much more prone to fog than the eastern side. This is usually because drier air is being carried slowly southwards on the eastern side; progressive drying inhibits fog formation. The reverse effect is found on the western side. Here the drift from the south brings progressively moister air. Even if the flow at the surface drops off to near zero the moisture continues to increase just above the surface and fog becomes much more likely.

Clearance of fog by spread of cloud

Although it may be practically calm near the ground there may be a wind of 10–20 knots at 2,000 feet. This can bring clouds over the foggy area. The arrival of a sheet of low cloud often clears away the fog. This may happen at any time, often long before dawn, especially if the fog is shallow.

Clouds lose heat by radiation, both upwards and downwards. The arrival of a sheet of cloud intercepts the outgoing radiation from the ground, absorbs it and radiates heat back to the ground (see fig. 137).

Persistent fogs

One often hears a BBC forecast suggesting that although it is foggy now the fog will clear during the morning. It is wise to treat such predictions with caution during the winter months, especially if you are planning to fly to a fog-bound airfield. Methods of fog prediction require precise details of the depth of fog and the temperatures through the fog. Such information is seldom available to the forecaster.

Some of the longest lasting winter fogs, formed some time after sunrise, persisted two or three days and finally cleared during the night when an approaching front brought cloud and stronger winds. Over the British Isles, northern Europe and the western part of Russia, winter fogs tend to persist if an anticyclone becomes stationary just to the east. The mountains and higher hills may be bathed in sunshine throughout the day but the sun is seldom strong enough to penetrate the layer of fog covering low ground.

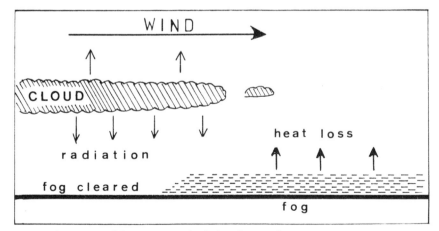

137. *Clearance of fog by spread of cloud over the top*

15

Weather maps and forecasts

How and where weather is observed

(a) Most surface observations come from airfields on land and merchant ships at sea. Reports also come from automatic weather stations, but these usually omit cloud and weather. At sea there are some 7,000 Voluntary Observing ships which are supplemented by 1,000 buoys; a few of these buoys are moored but most just drift with the ocean currents.

Hang-glider pilots have installed their own automatic stations at selected hilltop sites in England and Wales. The system is named 'Wendy Windblows'.

(b) Upper air soundings used to be made by aircraft but now come chiefly from radiosondes; these are small expendable transmitters carried aloft by balloons. They measure air pressure, temperature and humidity up to 53,000 feet and higher. Wind velocities are determined from the drift, which can be found by radar tracking or from internal radio navigation systems. There are about 400 radiosonde stations on land. Over the oceans, commercial aircraft make regular wind observations; the process can be automated and the data stored for interrogation by satellite.

(c) Satellites were originally just for cloud pictures, but now they also measure temperature profiles from sea level to the high stratosphere. The method combines infrared and microwave radiation at a number of wavelengths. A single orbit by an NOAA satellite can produce hundreds of data points; these are essential for mapping the empty oceans. Geostationary satellites measure upper winds by tracking the movement of clouds at half-hourly intervals. They also map the distribution of moisture.

(d) Lightning flashes can be located by radio at distances of several thousand miles. Originally they were found by triangulating the bearings from three or four direction-finding stations using the very low frequency of 10 kHz. Modern systems use the time differences instead of the bearings.

(e) Rainfall patterns and intensities can be mapped by a network of radar stations.

Exchange of information

Before WW2, most Met. data was sent by W/T (Wireless Telegraphy). Met. messages were sent in an internationally agreed code consisting of a series of five-figure groups. These are much more concise than plain-language text.

Some years ago, amateurs could plot their own charts using data broadcast by W/T, RTTY (Radio Teletype) and Radio facsimile. When computers came into use, they exchanged data at a far faster rate using cable and satellite links. Then Morse code ceased and most RTTY broadcasts were withdrawn. As a result, amateurs were cut out until the Internet started up. Old-style Met. messages can still be obtained (by special arrangement) from main Met. centres, but they are no longer free.

Weather maps

Up to the 1960s, practically all weather maps were plotted by hand from the coded messages. Surface observations were entered on an outline map using an international scheme of figures and symbols which are recognised worldwide. Figure 138 shows an example of a single observation. Figure 139 gives the basic weather symbols. A full list of symbols and their meaning is in Appendix 1.

Manual plotting is still used, chiefly for studying the weather over a limited area, but most charts are plotted by computer. This controls a mechanical plotter for large charts, and makes 35 mm photos which can be enlarged to give big prints or displays

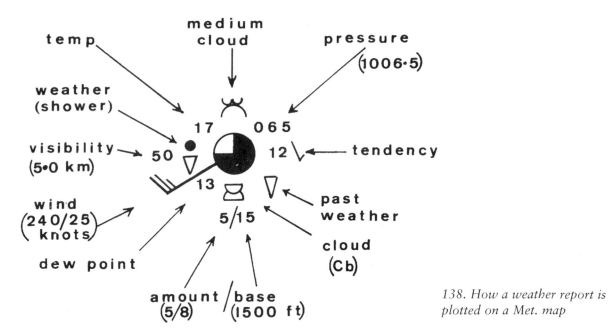

138. *How a weather report is plotted on a Met. map*

SYMBOLS FOR PRESENT WEATHER		
TYPES OF PRECIPITATION	OTHER ITEMS	
DRIZZLE	SMOKE	
FREEZING DRIZZLE	MIST	
DRIZZLE AND RAIN	FOG	
RAIN	FREEZING FOG	
FREEZING RAIN	SANDSTORM	
RAIN AND SNOW	BLOWING SNOW	
SNOW	RISING DUST	
RAIN SHOWERS	DUST DEVIL	
SNOW SHOWERS	SQUALL	
THUNDERSTORM	FUNNEL CLOUD	

139. *A table of symbols depicting present weather types*

them on a computer monitor. The monitor can combine different types of data in one picture and zoom in to enlarge details.

Chart analysis

In the old days, an assistant plotted the chart. The forecaster then analysed it, adding isobars and fronts in pencil. Now the computer both plots and analyses the charts; these cover the entire globe at many levels starting from sea level and extending up into the stratosphere. Years ago, the forecaster's problem was lack of data. Now there is far too much for one person to absorb, and computers have taken over most of the work.

Computer forecasts

These are also known as 'numerical forecasts'. This means converting all the variables of real weather into numbers which the computer can process. The answer is returned as an array of numbers which then have to be changed back into charts, graphs and tables.

The numerical process consists of several stages:

(1) Quality control. Before any calculations begin, a front-end computer has to check the incoming data to prevent false information from corrupting the forecast.

15 LAYER MODEL

140. *How the grids for a 15-layer model are superimposed*

(2) Transferring data to the grid. The human eye can make sense of bits of information, even when they are scattered irregularly over the chart, but most computers need regularly spaced data. The chart is covered with a grid like a fishing net whose knots represent grid points. Grid-point values are allotted by interpolating between the scattered observations.

Conditions change with height, so the machine needs several layers to represent the atmosphere in depth. Figure 140 illustrates the scheme for a 15-layer model. In this case, a 'model' is just a mathematical description of the weather processes to be calculated. The earliest models had just three layers; the latest use 30 or more.

(3) 4-D interpolation. Global charts are usually analysed at six-hourly intervals, but valuable satellite data and aircraft reports come in at all hours. A fourth dimension of time is added to the three spatial dimensions. Interpolation has to make allowance for the change and movement of weather with time. This involves making a brief forecast to convert or move the observation to fit the six-hourly standard.

(4) Smoothing. There are many different waves in the atmosphere; some of them can grow big enough to disrupt a numerical forecast. To get round this problem, the early models used 'filtered' equations which cannot predict these waves. However, the results were not satisfactory, so the programmers turned to 'primitive' equations. To prevent violent waves being set off by irregularities, the analysed data was adjusted to give a smooth surface which would not stimulate unwanted waves. Putting in new observations can be like dropping a rock into a pool; there is a big splash and a series of waves radiates outwards. To prevent this commotion, new information is inserted gradually. Short forecasts are run to ensure that any new humps or hollows are assimilated without a wave-provoking jolt.

(5) Integration: time-steps. A global forecast uses some 45,000 grid points arranged in 31 layers. To make a forecast, about 5 million simultaneous equations have to be solved. This cannot be done in a single jump. Instead the changes are worked out in a series of short steps, often only a few minutes long, known as 'time-steps'. It is like defining a curve as a series of very short straight lines. The shorter the lines, the better the curve looks, but the longer it takes to calculate.

At the end of time-step one, there is a forecast for (say) ten minutes ahead. This new data is the starting point for step two. After several hundred repetitions, the machine will have made a forecast for three days ahead. Some routine forecasts extend out to ten days. Climatological forecasts run for months, but their methods are different.

(6) Putting in the weather. The earliest machines could not handle moisture, so they were unable to predict clouds and rain. The later machines can calculate evaporation and condensation of moisture. From this, they work out the depth of cloud and how much rain will fall. Change of heat due to incoming and outgoing radiation is calculated at each time-step. This determines the daily rise and fall of temperature and hence the formation of cumulus by day and fog at night.

Some features are too small to be calculated explicitly; instead they are represented by parameters, which are constants which have to be found statistically. Stability is computed directly, but small features like Cu and Cb are given by a parameter.

(7) Presenting the forecast. This involves converting the mass of numbers into charts, graphs and tables of wind and temperature. The computer draws isobar and contour charts by interpolating between the grid-point values. Early models with coarse-mesh grids could give misleading results if a small but deep depression lay within a single large grid square (see figure 141). This problem has been much reduced by using smaller grid squares.

Upper winds are usually displayed either as a contour field and/or a series of wind arrows. The change of wind with height is shown on 'Spot Wind' charts. These give the wind velocity and air temperature at many levels above a single spot. They are printed in a series of boxes as shown in figure 142.

Humidity, cloud amounts, precipitation and the freezing level are usually shown on separate charts depending on the user's needs.

A great variety of special-interest charts and graphs can be supplied, ranging from wave heights for mariners to high-level trajectories for long-distance balloon pilots.

Errors and limitations

There are two main types of error. The first depends on the grid size. A coarse-mesh grid cannot resolve small systems, so important features such as

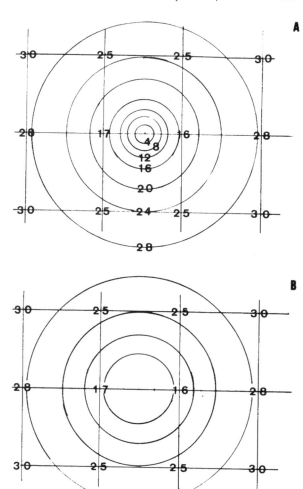

141. *How a coarse-mesh grid can lose a deep low:*
(a) actual analysis showing grid point values
(b) How computer draws the isobars to fit the grid point values but misses the depth of the low

thunderstorms may never be explicitly calculated by the model. The shape of mountains is smoothed out by a coarse mesh, so the drag they cause may be wrong. The problem can be reduced by using a very fine-mesh grid, but this raises new difficulties. If the entire globe is to be covered, the computation is extravagant. If only part of the globe is covered, the boundary regions will be corrupted by systems coming in from outside.

The difficulty can be resolved by using a 'nested grid'. A very fine-mesh grid is inserted ('nested') into the global grid. The global model keeps the boundaries updated while the fine details are being

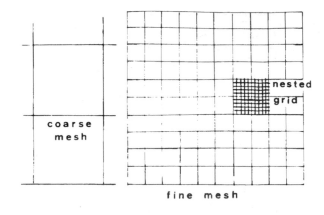

142. *Example of grids: coarse mesh, fine mesh and nested*

calculated inside the nested grid. The method is useful for intense features such as hurricanes and small systems such as sea-breeze fronts.

The second type of error has been called 'sensitive dependency'. If the weather is very sensitive to small changes, a variation of as little as one part in 1,000 may result in different forecasts coming from apparently identical conditions. For the first few hours, the two forecasts seem to keep in step, but then they gradually diverge. By day four they can be completely out of phase.

Many features of the weather cannot be measured with perfect precision. Often this does not matter, but occasionally the balance is critical. A boulder perched on top of a steep ridge might roll either way. Then a little push decides if a disastrous avalanche will fall in one valley or the other. There is no middle way. One way to find if initial errors are important is by making tiny changes to the input data and running the forecast again. A series of such forecasts is called an 'ensemble'.

Ensemble forecasts

Supercomputers are now so fast that a forecast can be repeated several times. Programmers arrange for tiny alterations to be inserted in the initial data each time the forecast is repeated. A series of repeats produces a set of forecasts known as an 'ensemble'.

Each forecast is different in some areas but hardly changes in others. The ensemble shows the regions where the forecast is particularly sensitive to errors. This can be particularly useful for forecasts which extend five to ten days ahead. The method can be adapted to show the most likely line of development. The probability can be given as a percentage.

Availability of forecasts

Met. centres have a wealth of actual and forecast information but a paucity of economical distribution channels. The best (but most expensive) way of getting weather information is to have a direct computer link with a Met. centre. A great variety of data can be obtained through the system known as MIST.

MIST (Meteorological Information Self-Briefing Terminal) is a PC-based system which links the user with the Central Forecasting Office in Bracknell. It can provide current weather reports and forecasts on a local, regional, national or global scale. The data can be arranged to suit the customer's needs.

For example, a flying club can be provided with TAFs, METARs, surface pressure charts, significant weather maps, wind fields, rainfall patterns and satellite images. Lightning displays can show strikes every five minutes. It is possible to 'zoom in' on particular areas and overlay geographical features on the display. Such elaborate weather data is not cheap; the price depends on the range of facilities required. It is cost-effective for airlines and large weather-sensitive companies but may be too extravagant for a private pilot.

PC displays of Met. forecasts are available from other European centres as well.

Speaking to a forecaster

The old days when anyone could ring up the nearest Met. Office and speak to a forecaster are long past. The number of forecast offices and weather centres is being reduced and the remaining ones have little spare time. Pilots are expected to use routine issues of standard forecasts which are available by phone, fax, PCs or the Internet. An aviation forecaster may still be consulted for special non-routine requirements, but the caller is expected to pay on-line using a credit-card number.

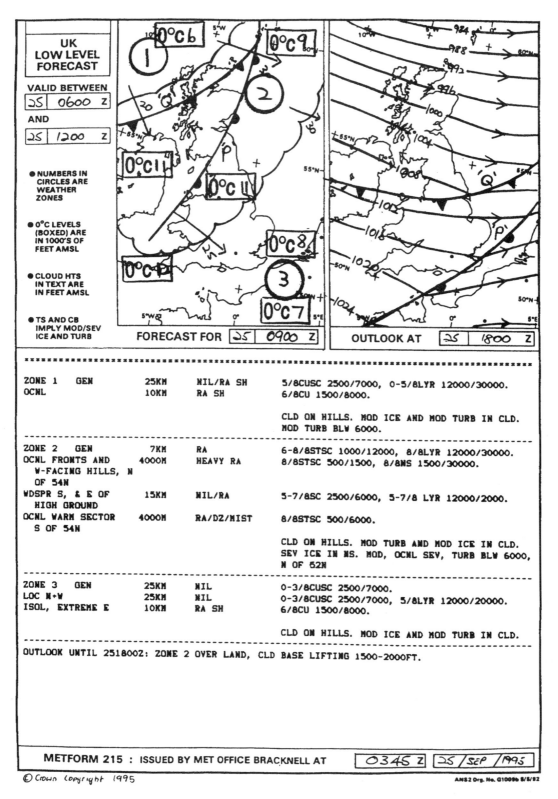

143. *Example of a low level weather chart received by MetFAX*

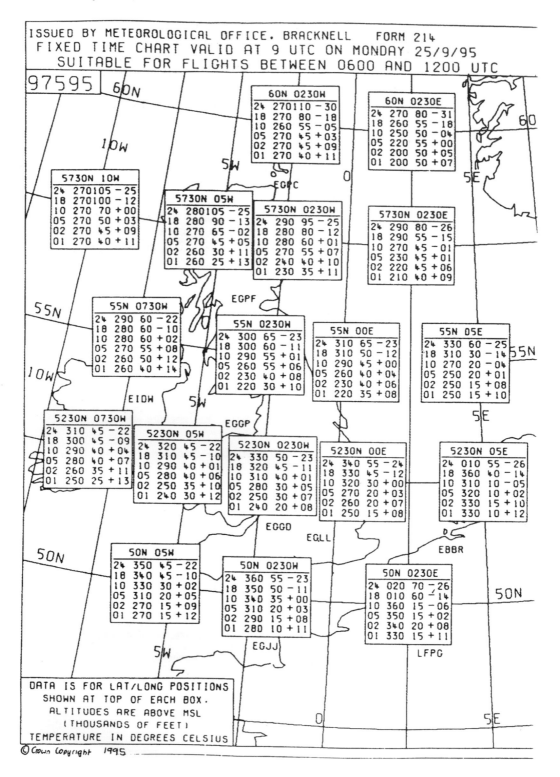

144. *Example of a spot wind chart received by MetFAX. (Both charts reproduced by permission of the Met. Office, Bracknell.)*

GET MET (Aviation Weather Services)

A comprehensive list of UK Met. products is contained in a small booklet entitled 'GET MET' which is available free from the Met. Office, Sutton House, London Road, Bracknell, Berkshire RG12 2SY.

Other services available

Dial-up fax (MetFAX) is a convenient way of obtaining forecasts and charts. Figures 143 and 144 show examples of MetFAX for aviation. They combine schematic charts divided into weather areas, a written forecast and a chart giving spot winds.

The menu of MetFAX products for aviation is obtained by dialling 0336-400-501 or 09060-700-501. This gives about 35 numbers for the various products. For example, the pages shown in Figures 143 and 144 were obtained by dialling 0336-400-503.

Internet service (MetWEB): the Met. Office has a web site at www.met-office.gov.uk which duplicates several of the MetFAX products. For e-mail, use metweb@meto.gov.uk

There is a charge made for most Met. services; however, free METARs and TAFs can be received by radio on both VHF and HF. HF transmissions require a single side-band receiver. VHF transmissions are for aircraft in flight and may be inaudible on the ground.

A list of world-wide radio weather services can be found in 'The Admiralty List of Radio Signals' published by the UK Hydrographic Office. There are several volumes, intended chiefly for mariners. Volume 3 part 1 contains details for Europe, Africa and Asia.

16

Satellites

There are two main types of Met. satellite: polar orbiting and geostationary. Most of those in polar orbit have been launched by the USA. The first few were the TIROS (Television and Infrared Scanning Orbiting Satellite). Later versions were launched by NOAA (The National Oceanographic and Atmospheric Administration); in 1999 the latest was NOAA–15. By the end of April 1999,

NOAA–15 had made some 5,000 orbits. The NOAA series have orbits inclined at about 99 degrees which takes them close to both poles; they are usually at heights of about 460 nm. Each orbit lasts about 102 minutes; there are 14 orbits a day scanning a swathe 1,400 nm wide. Each satellite obtains complete cover of the globe twice a day.

Other satellites have been launched by the

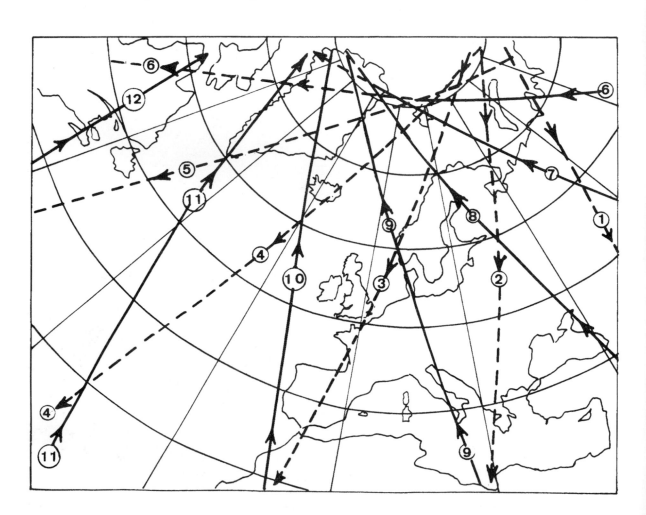

145. (a) Successive tracks of polar orbiting satellite NOAA–10

Russians. Their METEOR series operates at a height of 650 nm and takes 110 minutes to complete an orbit. Further Russian satellites are named OKEAN, SICH and RESURS but these do not send pictures regularly. The Chinese launched their FENG-YUN series and the Japanese plan to launch an ADEOS series. The European Met. Satellite Council hopes to put three satellites called METOP in polar orbit. Nearly all these transmit on frequencies between 137 and 138 MHz.

Pattern of NOAA scans

Figure 145(a) shows successive tracks of an NOAA satellite. Southbound tracks, numbered 1–5, are shown by pecked lines, while northbound tracks numbered 7–12 have full lines. Each track passes 25.5 degrees further west; this is the angle the earth rotates in 102 minutes. Figure 145(b) shows a single track with the sub-satellite position marked every minute since crossing the equator northbound.

Geostationary satellites

These use an orbit which keeps them over the equator. They operate at a height of 19,330 nm. At this level an orbit takes 24 hours, so the satellite matches the rotation of the earth and always remains over the same position. There are normally five geostationary satellites in operation daily; spares are kept in parking slots and can be moved into position to replace failures.

Meteosat is on the Greenwich meridian, Insat (for India) is near latitude 70°E, GMS is near 140°E, GOES-W is at 140°W and GOES-E is at 75°W. On 10 June 1997, the Chinese launched Fen Yun IIB, now located at 119°E.

Geostationary satellites are 'spin-stabilised'. It takes 25 minutes to make a complete scan of the visible globe; this is repeated every half hour so that a time-lapse film can be displayed. The original pictures from Meteosat are sent down to the ground station at Darmstadt in digital form. The full globe

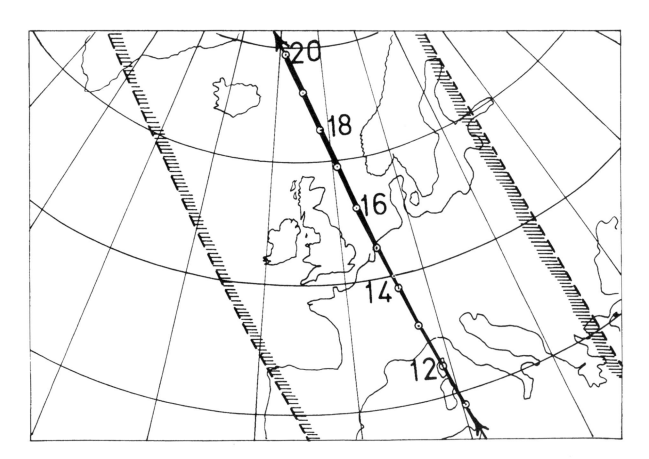

(b) sub-satellite position each minute after the equatorial crossing on a Northbound section of the orbit

image is then divided into sectors and sent back to the satellite as a series of enlarged images. These are rebroadcast in analogue form. The primary data is now encrypted; it costs about £650 to obtain the key to decryption.

Wavelengths scanned

The NOAA series scans the globe at five wavelengths. These are given below in microns. (A micron is a thousandth of a millimetre.)

NOAA series Channel 1 (Visual image) 0.6 microns
            Channel 2 (Visual)   0.9
            Channel 3 (Near IR)   3.7
            Channel 4 IR (infrared) 10.8
            Channel 5 IR   12.0

Meteosat uses 0.4–1.1 microns for visual images
            5.7–7.1 microns for water vapour
            10.5–12.5 microns for infrared.

Reception of data

The American orbiting satellites were designed for universal reception using their APT (Automatic Picture Transmission) system. Data is also recorded for playback to certain ground stations. Adequate VHF reception is usually possible using crossed dipoles or a quadrifilar aerial. For HRPT (High Resolution Picture Transmission) one needs a steerable aerial (usually a parabolic dish) rotated by an autotrack system. Geostationary satellites such as Meteosat can be picked up with a long Yagi aerial or by using a dish.

Some radio amateurs have built their own sets, but complete commercial systems are available; the cost depends on the standard of facilities built in. Many systems save money by using a PC to process and display the images on a monitor.

Future developments

A cloud-profiling radar is planned for CLOUDSAT to be launched by NASA in 2003. This is intended to study the 3-D structure of clouds; present systems only see the cloud tops.

Interpreting the images

(1) Visual images depend on reflected light, and look like black and white photographs. Ice and clouds are white, the land has various shades of grey and the sea looks black. A field of small, closely spaced cumulus clouds may appear grey if the clouds are too small to be resolved individually.

(2) IR images depend on the emission of radiation; this varies with temperature. High clouds whose tops are very cold appear white while warm areas such as tropical oceans look black. With some systems, the user may specify a range of false colours rather than black and white.

Land temperature varies between day and night, but cloud temperatures change little, so the IR image looks much the same by day or night. Fog and very low clouds have almost the same temperature as the underlying surface and can be difficult to detect using IR. Figures 146 and 147 show North Sea fog which shows up clearly in the visual picture but is lost on the IR image.

(3) The near-IR images which use a wavelength between the visual and IR are useful for distinguishing between fog and the underlying ground or sea.

Using satellite pictures

Individual pictures can be difficult to interpret by themselves; they are easier to understand when viewed beside a weather map which shows highs and lows and the long, curving lines of fronts. It helps to be able to compare IR and visual pictures. Changing from one to the other makes a sheet of high clouds stand out from a layer of low stratocumulus. The cold summits of Cb extending high above the other clouds are prominent features of IR images. On visual pictures, Cb may be hard to distinguish unless they cast a shadow. Software exists to convert the flat-looking image into a sort of 3-D picture with tall clouds sticking up like mountains.

Time-lapse pictures

A succession of half-hourly pictures from a geostationary satellite can be stored and played back as a time-lapse film. This gives a much better impression of the dynamics of the weather. Each frame uses a lot of memory, so a long series of half-hourly pictures can fill up all the space available on a PC.

146. and 147. North Sea fog, a comparison of visual and infra-red images

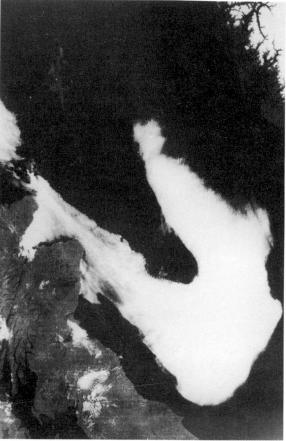

146. Infra-red picture of sea fog in the North Sea. The warm land shows up black on infra-red; the sea temperature and the fog temperatures are too close to be clearly distinguished (Picture from Dundee University)

147. Visual picture of North Sea fog. The fog shows up clearly against the sea which appears black on this visual image. The snow-covered mountains of Norway are in the top right-hand corner (Picture from Dundee University)

Correction for distortion

Pictures from geostationary satellites are distorted by the curvature of the earth. The distortion increases with latitude. The pictures can be reprocessed to eliminate much of the distortion, but small details cannot be resolved.

Frontal activity seen on IR

When a front becomes more active, the associated high cloud (mostly cirrostratus) usually becomes higher and more extensive. The band of very white cloud on the IR image is carried downwind. Streamers of cirrus fan out, giving a rather 'hairy' appearance. This sometimes indicates a deepening system.

When fronts start to weaken, the band of cirrus grows narrower, breaks appear and eventually the air aloft starts to subside. The very white clouds then disperse, leaving only low, grey-looking clouds.

Effect of anticyclones and ridges

The subsidence of air above a surface anticyclone makes most of the cloud disperse. Large, cloud-free areas often develop over highs and near the axis of ridges. However, fog may obscure the surface.

Lee waves

Waves over and to the lee of mountains are marked by parallel bars of cloud with clear slots in between. The pattern is similar to the ripples left on a flat sandy beach by the retreating tide. Cirrus cloud does not readily form bars; instead it just streams away in a solid sheet. The upwind end stays anchored to the axis of the main ridge until the wave collapses. Strong wind shear in a jet stream can cause shear waves at the cloud top. Unlike lee waves, these are not stationary. They appear as striations at right angles to the flow.

Life of a frontal low

Many depressions form on a slow-moving front. The life history is shown in figure 23 (Chapter 2). The series below shows examples of various stages seen on IR satellite pictures provided by Dundee University.

Figure 148 shows an open wave with a well-marked bulge on its northern edge. The dense cirrus

148. *Frontal wave showing bulge on northern edge of cirrostratus and transverse striations*

shows up very white with a sharply defined northern edge. This is where one would expect to find the jet stream blowing almost parallel to the cirrus edge. There is generally a strong wind shear near the jet which can cause transverse waves to form. These show up as narrow bars or closely spaced striations where the cloud sheet breaks up on the southern flank. They are often a sign of high-level turbulence.

Figure 149 is a more complicated picture. An old low had formed a whirl at the top left, but the important feature is the hook-shaped cloud near the centre. The hook marks where the frontal system is becoming occluded; the occlusion is beginning to curl back round the centre of the low.

149. Hook cloud developing over a deepening depression

Figure 150 shows a more pronounced hook with an almost cloud-free slot pushing in south of the centre. This kind of slot marks the intrusion of a tongue of dry air which is subsiding from high levels. It often indicates a vigorous low which is slowing down. Gales are common on the southern flank.

150. Hook cloud with a clear slot of dry air being drawn into the circulation

Figure 151 illustrates an older low where the dry slot has been wound up into the circulation west of Ireland to form part of a slow-moving vortex. Over Ireland the cold frontal cloud was clearing from the west. The rear edge of the frontal cloud extends southwards towards Lisbon. Here it is being overtaken by a band of jet-stream cirrus running eastwards from a new wave depression in the SW corner of the picture. Once again the jet stream has produced transverse striations indicating high-level turbulence over and just west of Portugal. The new frontal wave is shown by a thickening of the cirrostratus with a curved northern edge (rather like in figure 148).

151. Dry slot being wound into the circulation of a vortex with a new frontal wave developing to the SW

Figures 152 and 153 show the latter stages of a filling vortex off the west coast of Ireland. The pictures were taken on successive days with an interval of 20 hours between them. In figure 152, the curl of dry air can still be seen winding round a deep low of 964 mbars. In figure 153, the low had started to fill up and the cloud vortex was degenerating into a confused swirl.

152. *Later stage in a filling cloud vortex over an old depression W of Ireland*

153. *Taken 20 hours after figure 152, showing the cloud vortex losing definition as filling continued*

Cloud streets and vortex shedding

Cumulus which forms under an inversion is often aligned into long streets parallel to the wind. Spectacular streets form when Arctic air streams out over the Atlantic from the Davis Strait (figure 154). In rather similar conditions but with lighter winds, a pattern known as 'vortex shedding' appears to the lee of isolated mountainous islands such as Jan Mayen (figure 155). The airflow curls first round one side of the island, then breaks off and curls round the other side. A series of eddies then blows away downstream.

The same effect on a much smaller scale occurs when the air blows round a tall cylindrical chimney. The flow breaks away as a series of vortices which (in extreme cases) can make the chimney vibrate.

154. *Cloud streets extending from the Davis Strait as Arctic air spreads into the Atlantic*

155. *Vortex shedding. Contra-rotating vortices trailing away from the mountainous peak of Jan Mayen Island in the Arctic*

17

Gliding weather

Cross-country flights in thermals: conditions required

Thermals should:

(a) start early, continue throughout the day without interruption and only die out towards sunset. The route should be free from incursion of air behind a sea-breeze front which would suppress usable thermals.

(b) in most cases, be marked by small cumulus with a high cloud base, heavy Cu with wide gaps are a handicap especially if cloud flying is not permitted.

(c) feature no big showers; if there are any at all they should be small, brief and widely scattered. Areas of Sc formed by spreading out of Cu are most unwelcome.

(d) winds should be light with little or no vertical wind shear to distort thermals (stronger winds are only useful for down-wind flights or ridge flying). Speeds of 5–15 knots are usually best. Anything over 25 knots makes most closed-circuit flying difficult although up and down wind routes using cloud streets are still possible with speeds of 35 knots.

(e) the visibility should be good (more than 10 miles). Thermals are often weaker in hazy areas and poor visibility prevents pilots from assessing the clouds ahead.

The advantage of fresh cold air

Most of the conditions needed are found when a fresh supply of cold air spreads over. After a cold front has passed the air is cooler and less heat from the sun is needed to set off thermals. Hence, thermals start earlier in the morning and, if there is a continual supply of progressively colder air, persist later in the afternoon. The main problem is the depth of instability in the cold air mass. Deep instability leads to heavy Cu and Cb followed by showers or even thunderstorms.

Soaring conditions related to the pressure pattern

Figure 156 illustrates an anticyclone with its associated ridge lying between two frontal systems. Shading shows the area covered by thick cloud. (a) shows the plan view with remarks on cloud and weather. The dotted line marks the position of the cross-sectional view in (b). The best conditions are almost always found when a ridge of high pressure develops after the passage of a cold front. The weather between two frontal systems generally varies as follows:

(a) regions close behind a depression are almost always bad for soaring either because Cb are likely to produce showers or because the remains of old frontal cloud are wrapped round the centre.

(b) any small troughs of low pressure swinging round behind the old depression are also bad because they produce more showers and broad bands of thick cloud too. Cyclonic curvature to the isobars is a poor sign even if no definite trough line appears.

(c) isobars which are straight or only slightly cyclonically curved may favour development of cumulus streets. Streets are often associated with stronger-than-average winds which are a handicap for cross-wind legs.

(d) slight anticyclonic curvature (as in a weak ridge) is a hopeful sign. It often indicates slow subsidence aloft. This produces a stable layer which restricts the growth of large Cu. Provided that the winds are light, weak ridges often give good soaring weather. However, it is important that the air should be fairly dry. Too much moisture leads to spreading out of Cu producing large areas of Sc. This cuts off the sun and makes thermals few and far between.

(e) a strong ridge with marked anticyclonic curvature is nearly always a very good sign. It usually indicates considerable subsidence. This produces an inversion or stable layer which severely restricts the growth of Cu. Provided that the inversion is at least 5,000 feet (above the average level of fairly flat ground) there should be good soaring conditions. Since Cu will be small they will probably be closely spaced and are unlikely to produce enough shadow to restrict thermals.

(f) anticyclones almost always have strong subsidence inversions. This is not always an advantage; the base of the inversion may descend below the condensation level. As a result many summer anticyclones tend to become cloudless and the depth of blue thermals is too shallow for good soaring. In the hottest period of summer thermals may lift the subsidence inversion a thousand feet or more during the day, thus improving soaring in the afternoon.

Subsidence is greatest when the high is forming. Old highs which have begun to loose pressure have much less subsidence to maintain the inversion. After the anticyclone has persisted a few days the original inversion may sink so low that the sun's heating 'warms out' the lowest layer and removes the inversion. Soaring then improves greatly with thermals rising higher and being once again capped by small Cu.

Cloud cover. The importance of land track and shelter

The size of Cu depends on the depth of the unstable layer but the amount depends on how moist the air is. In the United Kingdom and north-west Europe, where winds often come in from the Atlantic or North Sea, the coastal regions tend to have more cloud with lower bases than areas well inland. The longer the land track the better conditions become. A range of hills is often very effective in drying out the moist air from the sea. Provided that no trough develops to concentrate the low-level moisture the best soaring will be found well inland and in regions sheltered by a range of coastal hills.

Cloud base

If the air has a low moisture content the dew point will also be low. Low dew points generally give the best soaring conditions. The base of Cu can be found (with sufficient accuracy) by taking the difference

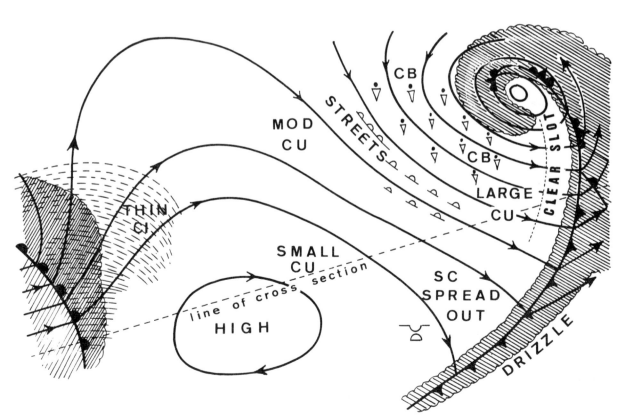

156. (a) Map of typical weather types when a high develops between two frontal systems

between the air temperature and dew point (in degrees C) and multiplying by 400. This gives the base in feet. For example, if you hear a VOLMET broadcast giving 'temperature 18, dew point 8', then the difference of 10 degrees implies a condensation level of about 4,000 feet. While the temperature is rising this will also be the approximate base of any Cu.

If you listen to VOLMET broadcasts be wary of the term CAVOK (pronounced 'Cav-OK'). It means Ceiling And Visibility OK. This term is used if the visibility is 10 km or more and the cloud base is above 5,000 feet. CAVOK sometimes means it is cloudless but there may be a complete cover above 5,000 feet.

Dew points and good soaring days

Good cross-country days are those with a cloud base of at least 4,000 feet by noon or early afternoon; this means that the air temperature and dew point must be well-separated. Over England the difference between the maximum temperature and the afternoon dew point is more than 10°C on good days. One can make a rough estimate of how dry the air will be from the forecasts of night minimum and day maximum temperatures on TV. If the wind is light the minimum temperature on clear nights depends on how dry the air is. The temperature falls more when the air is dry. The forecast minimum temperature for the night and the maximum for the following day are often a good guide to the dew point depression during the afternoon. Thus, a predicted night minimum of 2 followed by a day maximum of 15, giving a range of 13 degrees, is a very hopeful sign. Naturally it is no use if the rise in temperature is due to an approaching warm front.

The table below shows the range of dew point depressions observed over the United Kingdom on cross-country days.

Table 8. Dew point depression at the time of maximum temperature

Depression (Deg C)	Percentage of days
7–8	3%
9–10	9%
11–12	29%
13–14	25%
15–16	17%
17–18	10%
19–20	4%
21–22	3%

Note: The small percentage of occasions with dew point depressions of more than 18 degrees is typical of a predominantly maritime climate such as the United Kingdom and north-west Europe. Over a hot desert region or high plateau much drier air is common.

Spreading out of Cu to form Sc

This is one of the commonest reasons for a spoiled soaring day. It usually occurs when:

(a) there is an inversion which keeps all the Cu tops down to the same level;

(b) the air below the inversion is very unstable (so that cumulus forms with little heating);

(c) it is also very moist so that the base of the cloud is at least 2,000 feet below the inversion.

There are huge areas of this type of cloud over the

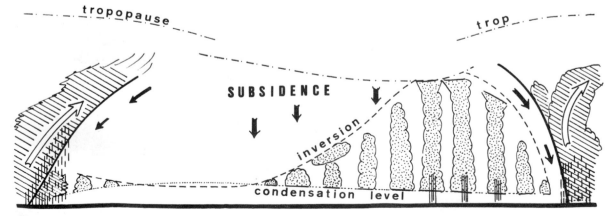

(b) *Cross-section view of cloud structure*

oceans on the eastern side of large anticyclones. Such cloud sheets tend to break up after the air has had a long track over the land, or been forced to climb over a range of coastal hills. Thus, the worst effect of spread-out occurs near windward coasts.

Dispersal

If the cloud sheet cuts off the sun completely the thermals usually die out. Then absence of thermals to maintain the cloud results in the sheet eventually breaking up. However, soon after the sun has broken through the thermals will start again and carry more moisture aloft to rebuild the cloud sheet. The factors which lead to dispersal are:

(a) The arrival of a strong ridge or high. The subsidence usually brings the inversion down until it is close to the condensation level; this makes the Sc layer too thin to persist. As each thermal rises up and bumps against the inversion the disturbance brings some warm dry air down through rifts in the inversion. This disperses the Sc layer.

(b) Falling pressure ahead of a trough. When a trough approaches there is usually large-scale ascent of air. This lifts and breaks the inversion so that Cu no longer stop at a fixed level. The Sc sheet then breaks up but the cumulus may grow too large for good soaring.

The effect of upper cloud

High cloud may spread more than 500 miles ahead of an advancing warm front. Non-frontal troughs are often preceded by a hundred-mile belt of upper cloud. The arrival of such cloud nearly always spoils the thermals.

(a) **Cirrus** Thin cirrus slightly reduces the strength of the sunshine and makes thermals less vigorous. Thickening Cs ahead of a warm front or trough has a more serious effect on thermals. The effect depends on the time of arrival. Cs arriving before the ground has warmed up will often prevent any thermals forming unless the air is very unstable. Cs which arrives during the afternoon when thermals are already well developed and Cu are all active is much less serious. It nearly always reduces the strength and frequency of thermals but only stops them if they were shallow and weak in the first place.

(b) **Altostratus** nearly always brings thermal activity

to an end; there are some exceptions when the air is very unstable and the cloud arrives when the ground has already become warm. Then weak and usually very smooth thermals can be found, even under cloud thick enough to obscure the sun. A few of these declining thermals may last until the arrival of rain.

The influence of soil moisture: wet ground often means weak thermals

Over a fertile countryside a large part of the available heat is used just to evaporate surface moisture or the water transpired by growing vegetation. When a lot of heat is used in evaporation there is less remaining to set off thermals. As a result dry hills produce more and stronger thermals than damp valleys. During and immediately after a wet spell thermals are generally weaker and the cloud base is lower. In the United Kingdom and much of north-west Europe really good soaring days are unlikely on the day following rain.

Wind speed

The cross-country speed of modern sailplanes is now fast enough to make the wind speed less critical than it was in the 1950s and 1960s. In those days almost all long flights were made down wind. Even now the wind speed remains a major factor in flight planning. This is particularly so for less experienced pilots who have not yet learned how to fly as efficiently as top-class competition pilots. The table below is based on closed circuit flights made in aircraft ranging from older wooden gliders to more modern GRP aircraft.

Table 9. Wind speeds on closed circuit days (flights of 200 km or more)

Range of speeds (knots)	Percentage of days
1–5	15%
6–10	31%
11–15	25%
16–20	19%
21–25	7%
26–30	1%
31–35	—
36–40	1% (using cloud street and wave)

Notice the sharp drop in the percentage of successful flights when the wind exceeded 20 knots. For competition pilots in Open Class gliders the limit is significantly higher.

This table excludes flights made almost exclusively along mountain ranges where thermal lift did not play a major part. Such out-and-return flights have been achieved in much stronger winds.

Wind direction

The wind direction is important only as a guide to the previous history of the air. The trajectory (track followed by the air) is a better indicator of soaring conditions than the actual wind direction on the day of the flight. The table below shows the percentage of good soaring days for eight main directions.

Table 10. Frequency of good soaring days in England for different air trajectories

Average direction	Percentage of very good days
From the S	nil
From the SW	1%
From the W	17%
From the NW	18%
From the N	33%
From the NE	16%
From the E	6%
From the SE	nil
Still air	8% (mostly near a High)

Although this table only applies to flights in England the general distribution is similar in north-west Europe. The actual wind direction on the day of the flight does not give such a clear-cut distinction between directions. Air from the south may have come originally from a Northwesterly direction and followed a curving track to reach the flight area. Even so, few good soaring days were reported when Southerly winds were blowing.

Why Southerly airflow is seldom any good

Most flows from the south bring warm air which has been cooled at low levels by its passage over colder ground; this causes a temperature inversion which delays the development of thermals or restricts their height. Ground temperatures may rise far above normal without setting off good thermals. When the inversion is finally broken by high surface temperatures large Cb may develop leading to thunderstorms in the afternoon and evening. Thundery outbreaks are particularly likely in very warm southerlies if a cold front (even a very weak one) comes within two or three hundred miles of the area.

Watch out for the appearance of altocumulus castellanus. These clouds are often like puffy small cumulus at heights of ten to twenty thousand feet above the ground. They are not produced by thermals rising from the surface but by slow ascent of medium-level air. These clouds are a sign that a potentially unstable air mass is becoming actually unstable aloft. Thunderstorms usually follow within 24 hours, often within 12 hours.

Not all North winds give good soaring. Some subsidence is needed

Although a flow from the North gives the best chance of good soaring, the day will not give ideal conditions unless the air has subsided enough to prevent showers. A cyclonic Northerly is almost always too showery; straight Northerlies are better provided the air has had a track overland. Coastal regions are liable to excessive cloud cover, often with a rather low cloud base. Anticyclonic Northerlies are the ideal, especially in regions sheltered by mountains.

Spring Northeasterlies

Record-breaking flights have been made when very cold NE winds extend over Europe and the British Isles during the spring months of April and early May. The air comes from polar regions and is so cold that thermals develop very early in the day. Provided that the air has been dried out by passing over the mountains of Scandinavia conditions should be ideal. As usual, the best regions are those with anticyclonic curvature. Closed-circuit flights are best near the high-pressure region where winds are light. Further from the high the stronger winds favour downwind flights towards the SW. The general pattern is shown in fig. 157. The best conditions often occur a day before frontal cloud spreads in from the North or Northwest (not immediately after the cold front has passed).

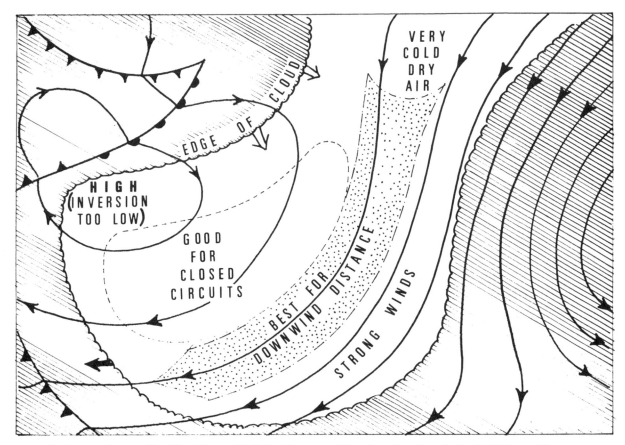

157. Spring northeasterlies showing typical cloud cover and regions for closed-circuit flights and downwind flights

Conditions for wave flying

The basic conditions for wave are:

(a) a wind of at least 15 knots blowing across a ridge (speeds of more than 20 knots may be needed for some mountains). The wind should be nearly at right-angles to the axis of the ridge. A difference of up to 30 degrees may not matter for a long ridge but if it is much more the low-level flow is liable to be deflected along the ridge rather than over the top. Cold or very stable air is often reluctant to go over the top and may be blocked by the high ground.

(b) there should be an inversion or very stable layer not far above the mountain tops with a deep layer of less stable air aloft.

(c) the wind speed should increase with height but the direction should remain fairly constant. An average increase of 1.5 knots/1,000 feet is usually sufficient but much stronger shears occur.

Average variation of wind speed with height

When all United Kingdom climbs of 10,000 feet or more were averaged the wind speeds increased as follows:

> 3,000 feet 28 knots
> 10,000 feet 37 knots
> 18,000 feet 50 knots
> 30,000 feet 69 knots

Changes of wind direction during the climb

In the majority of cases the change of wind between the bottom and top of a wave climb was less than 30 degrees. Sharp changes in wind direction are apt to result in the smooth wave flow breaking up into turbulence.

Table 11. Percentage frequency of changes of wind direction between 3,000 feet and top of climb

Direction change	Frequency (percentage)
less than 10 deg	33%
10–15 deg	32%
20–25 deg	22%
30–35 deg	8%
40–45 deg	3%
50–55 deg	1%

Winds below 3,000 feet were not used in this survey because the direction in valleys is apt to be un-representative of the general flow aloft.

Table 12. How the height of the stable layer varied

Height range (feet)	Base	Top
Below 1800	16%	—
1800–3200	14%	—
3200–4800	27%	4%
4800–6400	26%	10%
6400–8100	12%	17%
8100–9900	2%	30%
9900–11800	1%	13%
11800–13800	—	14%
not definable	3%	13%

None of the mountains exceeded 4,400 feet and the majority of ridges were below 3,000 feet. The stable layer is normally found much higher for waves over Alpine-sized mountains. It may be seen that for United Kingdom waves the base of the stable layer was generally within 3,000 feet of the highest ground upwind.

Features on weather maps which suggest lee waves

(a) **Suitable wind directions:** This varies depending on the direction in which the main ridges lie. Over the middle of the Scottish Highlands waves have been observed with all wind directions but at most sites the waves only develop in winds which have crossed the mountains. For most of the British Isles winds between Southwest and Northwest are most likely to give waves. East wind waves have been observed on the Western side of high ground but great heights are

rare with East winds. This is because it is less common to have the necessary increase of wind speed with height.

(b) **Influence of major centres of high and low pressure:** Soarable waves often occur over a wide band centred roughly half-way between the centres of high and low pressure. Close to a low there is normally too much thick cloud, the air is often too unstable and the wind direction is not constant with height. Near a high the low-level winds are too light.

(c) **Effect of fronts:** The frontal surface is nearly always marked by a very stable layer. The thermal winds blow almost parallel to the line of the front so that if the front lies almost parallel to the isobars the thermal wind will also be parallel to the isobars. In these circumstances the wind direction will change little with height but the speed will increase upwards. Thus, a slow-moving front meets most of the requirements for lee waves.

Satellite pictures almost always show lee waves in the upper clouds where a slow-moving front lies at right-angles to a major range of mountains. Sail-planes cannot always reach these waves because the weather is too bad at low levels. However, where the front lies across a ridge in the pattern of isobars there is a fair chance that the front will be weak and the cloud layers thin enough to break on the lee side of the mountains (see fig. 158). When such a slow-moving front lies at right-angles to a long range of hills conditions can be good for long cross-country flights. Even in small countries such as the United Kingdom flights of 500 km are possible. Much longer flights have been achieved to lee of major mountain ranges. The lee waves can be found both on the warm and the cold side of the front. Waves on the cold side are usually more extensive.

Jet streams

Jet streams are usually found on the cold side of a surface front. The very strong winds associated with jet streams trap wave energy radiating out from the mountains and channel it down a sort of duct between the jet stream and the ground. These trapped waves extend a long way downwind, sometimes for hundreds of miles; this makes them good for cross-country flights. Another advantage is that the wave pattern is more regular so that a pilot can anticipate where the next lift will be found. The lee wavelength

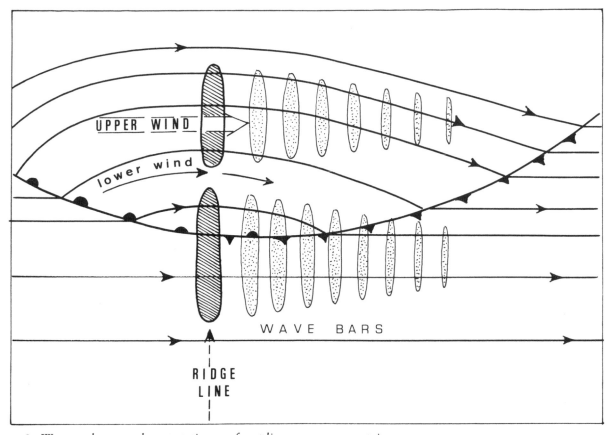

158. Waves where an almost stationary front lies across a mountain range

is usually fairly short near and on the warm side of the front. As one flies away from the surface front towards the jet stream the wave length becomes longer.

The best soaring area usually lies about half-way between the surface front and the axis of the jet stream (see fig. 159). Since the jet axis is on the cold side of a front it usually arrives overhead several hundred miles ahead of an advancing warm front. It also follows the passage of a cold front, but in this case the jet axis may be rather closer to the surface front.

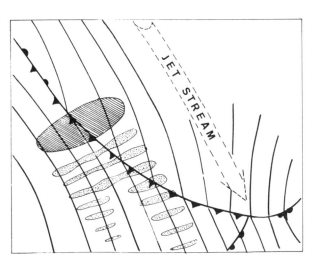

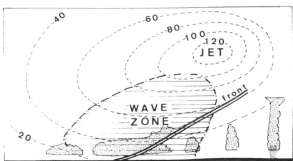

159. Cross-section showing sloping front, jet stream and the probable zone of lee wave activity

Glossary of meteorological terms

Accretion: In Met., refers to the growth of ice particles by collision and coalescence with water drops.

Adiabatic: A thermodynamic process in which heat neither enters nor leaves the system. The *Dry Adiabatic Lapse Rate* (*DALR*) is a constant 9.8°C per km (3°C per thousand feet).

Advection: Transfer of air mass properties by motion (usually horizontal), e.g. *Advection Fog* is formed when relatively warm moist air is carried over a cooler surface.

Aerological Diagram: A graph on which the readings of pressure, temperature and humidity can be plotted. It is used for displaying the results of vertical soundings made by aircraft or radiosonde and for calculating the effect of ascent or descent of the air mass. (See also *Tephigram*.)

Ageostrophic Wind: The vector difference between the actual wind and the geostrophic wind.

Air Mass: A large body of air, usually extending for several hundred miles, in which horizontal changes of temperature are small. It is often separated from an adjacent air mass by a *Front*.

Altocumulus (Ac): A layer cloud usually found at levels between 7,000 and 23,000 feet consisting of separate elements, often with humped tops, sometimes resembling high-level cumulus.

Altostratus (As): A smooth layer of cloud of uniform appearance usually covering much of the sky. When it is thin, the sun can be seen dimly as through ground glass.

Anabatic Wind: A local wind blowing up a slope, usually when the slope is heated by the sun. (See *Katabatic*, the reverse effect.)

Anafront: A front where warm air is ascending over the cold air.

Analogue: A past situation which resembles the current one over a large area. A forecast can be based on the assumption that the weather will develop in the same way as on the previous occasion. It is not a reliable guide; so far, no-one has found a good analogue on the global scale.

Anemograph: An instrument for recording wind speed and (usually) wind direction.

Anemometer: Instrument for measuring wind speed.

Anoxia: Total absence of oxygen (see also *Hypoxia*).

Anticyclone: A region of relatively high pressure shown on the chart by a system of closed isobars.

Anvil Cloud: The top of a cloud which spreads out to form a projecting wedge like an anvil; often seen extending from the tops of Cb.

APT: *Automatic Picture Transmission*. A method of broadcasting cloud pictures from satellites.

Arctic Sea Smoke: Also called evaporation fog, frost smoke, steam fog. It usually occurs when very cold air moves from frozen land out over open water.

Backing: Wind direction changing in a counter-clockwise direction, e.g. changing from west to south.

Bar: A unit of atmospheric pressure: 29.53 inches or 750.062 mm of mercury at the standard conditions of 0°C and gravity of 9.80665 m/sec/sec. 1 bar = 1,000 millibar or 1,000 hectopascals.

Barograph: A recording barometer.

Barometer: Instrument for measuring atmospheric pressure.

Barometric Tendency: The change of pressure during a specified period (usually 3 hours).

Beaufort Scale: A wind scale originally used by Admiral Beaufort for sailing ships but now chiefly confined to shipping forecasts.

No.	Speed	No.	Speed
0	less than 1 knot	7	28–33 knots
1	1–3 knots	8	34–40 knots (full gale)
2	4–6 knots	9	41–47 knots (severe gale)
3	7–10 knots	10	48–55 knots (storm force)
4	11–16 knots	11	56–63 knots (severe storm)

5 17–21 knots   12 64 and over (hurricane)
6 22–27 knots

Benard Cell: A pattern of convection in the form of hexagonal cells set up by gentle surface heating. It is sometimes seen in stratocumulus sheets where maritime convection is limited by an inversion. Air rises in the middle of the cell and sinks at the cloud-free boundaries.

Billow Clouds: Parallel rolls of cloud, usually aligned across the wind when there is vertical wind shear. They look like very closely spaced wave clouds but do not remain stationary.

Blizzard: Snow carried by a strong to gale-force wind. In the UK it means moderate to heavy snow with winds of 28 knots or more, causing drifting and a visibility of 200 m or less. In the USA it implies a temperature of –7°C or less with 30-knot winds (or more) and visibility 150 m or less.

Blocking: A term used for situations when the usual eastward movement of depressions is interrupted by a persistent anticyclone.

Bomb: An American term for an explosively deepening depression (12 mbar or more in 12 hours). Central pressure may fall 70 mbar in 48 hours.

Boundary Layer: (a) The layer of air where the wind speed is reduced by friction near the ground, often assumed to be 2,000 feet, and (b) the layer of air disturbed by convective currents.

Brocken Spectre: The coloured ring of light surrounding the shadow of an aircraft on the cloud top (originally observed from the peak of the Brocken mountain in Germany).

Buys Ballot's Law: A rule stated by Buys Ballot (1857); in the northern hemisphere an observer standing with his back to the wind has low pressure to his left and high pressure to the right. (The reverse is true in the southern hemisphere.)

CAPE: Abbreviation for *Convectively Available Potential Energy*. A measure of the energy released when convective clouds grow. (Approximately represented by the net positive area on a tephigram between the environment curve and the SALR from base to top of the cloud.)

Castellanus: Cloud in the form of turrets; cumulus-type clouds whose height is greater than their width.

CAT: An abbreviation for *Clear Air Turbulence*.

CAVOK: An abbreviation for *Ceiling And Visibility OK*. Means no precipitation, no cloud below 5,000 feet and visibility of 10 km or more.

Ceiling: (a) the maximum height attainable by an aircraft or a thermal, and (b) in the USA the lowest base of cloud which amounts to ⅛ or more.

Cell: A term applied to cumulus clouds usually referring to an individual cloud and its circulation. Thunderstorms may contain a number of cells (multicellular) or a single *supercell*.

Celsius: A temperature scale with the freezing point of water at 0 degrees and the boiling point at 100 degrees. (Exactly the same as Centigrade.) See also Kelvin.

Chaos: In Met., the developments in a dynamical system in which most orbits have sensitive dependence. When this occurs, tiny changes in the initial data cause huge variations in the final predictions.

Chinook: A warm, dry, Foehn-type westerly wind which occurs east of the Canadian Rockies, usually associated with lee waves.

Cirrocumulus (Cc): Cirrus cloud marked by ripples or tiny globules.

Cirrostratus (Cs): A sheet of cirrus cloud often producing a halo round the sun or moon.

Cirrus (Ci): Ice crystal clouds often appearing as fine hair-like streaks.

Cloud Street: A long line of cumulus aligned almost parallel to the wind.

Coalescence: Collision of small drops of water to produce larger drops. Also used when ice particles grow by collision with water drops.

Col: The saddle between two anticyclones and two depressions. A region of very light variable winds.

Cold Front: A front which moves so that warm air is replaced by cold.

Cold Front Wave: A wave-like undulation on a long cold front marking the formation of a new (secondary) depression.

Condensation Level: The level at which moisture in rising air starts to condense and form droplets of cloud.

Condensation Nuclei: Minute solid particles with a radius of greater than 1 micron needed to enable water vapour to condense into droplets after reaching saturation point.

Condensation Trail (often abbreviated to **Contrail**): A thin trail of ice crystals or water droplets produced when water vapour from an aircraft engine cools below the condensation temperature. Contrails are usually seen behind high-flying aircraft. They may persist and grow much wider when there is also cirrus cloud, for example ahead of a warm front.

Confluence: The flowing together of adjacent streamlines. (The reverse is termed **Diffluence.**)

Contours: On a Met. chart, contours are lines of constant height showing the slope of a pressure surface. They are used to find the geostrophic wind at upper levels.

Convection (in the atmosphere): Transfer of heat by the motion of a substantial volume of air.

Convergence: The coming-together of currents to cause an accumulation of air. (The opposite is termed **Divergence.**)

Coriolis Force: An apparent acceleration imposed on air moving over a rotating earth. The horizontal component acts to turn air to the right in the northern hemisphere (to the left south of the equator). When the force due to the pressure gradient is exactly balanced by the coriolis force, the wind is geostrophic and blows parallel to the isobars.

Corona: A series of coloured rings round the sun or moon caused by diffraction of light by water drops. It differs from a halo which is larger and caused by ice crystals.

Cumulonimbus (Cb): A very large cumulus cloud from which showers of rain, snow or hail may fall. It may develop into a thunderstorm and can produce squalls, occasionally tornadoes or waterspouts.

Cumulus (Cu): Clouds which grow upwards to form bulging domes or cauliflower tops, usually produced by strong upcurrents called **Thermals.**

Cyclogenesis: The formation or deepening of a cyclone (**Depression**).

Cyclone: A region of low pressure surrounded by a series of closed isobars. The term is usually restricted to lows in tropical latitudes. Elsewhere the term **Depression** is used.

Cyclonic Circulation: Airflow which curves round anti-clockwise in the northern hemisphere and clockwise in the southern hemisphere.

Deepening: A term used when the pressure at the centre of a low decreases with time. (The opposite is termed **Filling.**)

Density: Mass per unit volume.

Density Current: An almost horizontal pattern of flow due to differences of density in air or liquids. (See also **Gravity Current.**)

Depression: A region of low pressure surrounded by a series of closed isobars.

Deterministic: Later states evolve from earlier ones according to a fixed law usually expressed as an equation.

Dew: Condensation of water vapour on a cooled surface.

Dew Point: The temperature to which air must be cooled to become saturated with water vapour. Further cooling leads to the formation of fog or cloud droplets in the atmosphere or the deposit of dew on the surface.

Diffluence: The moving apart of adjacent streamlines. The reverse of **Confluence.**

Dissipation Trail: (often shortened to **Distrail**): The opposite of a **Contrail**, occasionally seen when an aircraft flies through a very thin cloud layer. The cloud dissolves where the wing-tip vortices produce downward motion.

Diurnal Variation: Changes which occur during the day such as the rise and fall of temperature or the regular variations of pressure.

Divergence: The moving apart of air currents to cause a depletion of the air. (The reverse of **Convergence.**) Divergence at one level is often almost balanced by convergence at another. When such balance is not achieved, the surface pressure rises or falls.

Doppler Radar: Radar which detects motion towards or away from the set by measuring the change in frequency between outgoing and reflected signals.

Downburst: A very powerful descent of air under or near a Cb. Vertical velocity may exceed 60 knots. Damage may be caused when the downburst reaches the ground and spreads out horizontally. (See also **Microburst**, which is a small downburst.)

Drizzle: Drops of water with a diameter between 0.2 and 0.5 mm. It is usually caused by coalescence of droplets of cloud. It evaporates if it falls through dry air; drizzle at ground level indicates moist air below cloud and suggests a very low cloud base.

Dry Adiabatic: A line on an aerological diagram showing the *Dry Adiabatic Lapse Rate* (9.8°C per km or 3°C per thousand feet).

Dry- and Wet-Bulbs: A matched pair of thermometers, one of which has the mercury bulb covered with wet muslin. The difference between the temperatures enables the dew point and humidity to be calculated.

Dry Line: A term used chiefly in the USA for the division between warm moist air from the Gulf of Mexico and dry desert air. Severe thunderstorms may develop along this line.

Dust Devil: A whirlwind marked by a rapidly rotating column of dust. It is often observed over desert areas where intense solar heating causes high temperatures leading to strong upcurrents.

El Niño: Spanish name for The Christ Child. It refers to a reversal of the ocean currents bringing warm water to the usually cold coasts of Peru and Chile around Christmas time in some years. The warming affects much of the eastern tropical Pacific ocean and causes climatic changes over a much wider area of the globe. Normally dry regions get exceptional rainfall while wet areas may have droughts instead. Its influence extends from the Indian monsoon to rainfall in the southern Sahara.

Ensemble Forecasts: A set of numerical forecasts made to test for sensitivity to small errors in the initial data. After the original forecast has been run, tiny changes are made to the initial data and the forecast repeated. If the prediction is sensitive to small errors in the starting conditions, the successive forecasts will diverge from the original.

ENSO: Short for the El Niño Southern Oscillation.

Entrainment: The mixing of surrounding air into the updraft of a cumulus cloud, thus diluting the thermal while increasing its volume.

Entropy: A function of the temperature pressure and volume of the air. It is proportional to the logarithm of the potential temperature. A dry adiabatic on an aerological diagram is also a line of constant entropy.

Environmental Temperature: A term used for the measured air temperature over a considerable depth of the atmosphere.

Evaporation: The change of liquid water or ice into water vapour, a process which requires extra energy to effect the change of state. (See also *Latent Heat.*)

Evaporation Fog: See *Arctic Sea Smoke.*

Eye of the Storm: The central region of an intense tropical storm (hurricane or typhoon) where the wind falls light and a small hole appears in the clouds.

Fahrenheit: A temperature scale which has the freezing point of water at 32°F and the boiling point at 212°F. To convert to Celsius, subtract 32 and multiply by 5/9.

Fallstreak: Another name for *Virga*, the trails of precipitation falling from a cloud but not reaching the ground.

Fax: An abbreviation of facsimile, a method of sending charts or documents by radio or telephone line.

Floccus: Puffs of medium- or high-level cloud like small cumulus often with trails falling from it.

Foehn: A warm dry wind on the lee side of mountains usually associated with lee waves.

Foehn Gap: A break in the cloud sheet caused by descending air in a lee wave system.

Fog: Reduction of visibility due to suspended droplets of water, sometimes mixed with smoke. For aviation the visibility is less than 1 km. For the motoring public the visibility must be less than 200 yards.

Fog Point: The temperature at which fog forms when the air is cooled.

Fractal: A set of points whose dimension is not a whole number; patterns such as a coastline, mountain outlines and fern-like shapes can be fractals.

Fractus (fra.): Broken or ragged low clouds.

Freezing Fog: Supercooled droplets of fog which freeze on impact to form *Rime* ice.

Freezing Level: The lowest height above MSL where the air temperature is 0°C.

Friction Layer: The layer of air where the wind speed is reduced by surface friction. See also *Boundary Layer.*

Front: A transition zone separating air masses of different density (temperature). The frontal surface usually slopes at a very shallow angle (about 1:100) and in some cases may extend from the surface up to the base of the stratosphere.

Frontal Fog: Fog which forms at or near a surface front, usually caused by warm rain falling into colder air near the surface causing saturation there.

Frontogenesis: A term used when a front forms or becomes more active.

Frontolysis: The weakening or disappearance of a front.

Frost: Occurs when the air temperature at or near the ground is below the freezing point of water. The term also refers to the icy deposits on objects, e.g. *Glaze*, *Hoar Frost* and *Rime*.

Funnel Cloud: A funnel-shaped protuberance below the base of Cb; rapid rotation of air makes the pressure drop within the funnel forming a cone-shaped cloud. If the funnel descends to the surface, a tornado forms over land (waterspout over the sea).

Gale: A wind speed of Beaufort Force 8, mean speed 34–40 knots.

Geopotential Height: A modification of the true height to take account of the variation of gravity at the surface of the globe. If g = 9.8 m/sec^2, geopotential and true height are the same. When the value of 'g' is increased, the geopotential height is greater.

Geostrophic Departure: The vector difference between the actual wind and the geostrophic wind.

Geostrophic Wind: A horizontal wind blowing parallel to the isobars when the pressure gradient is exactly balanced by the *Coriolis Force*.

Glaze: A transparent layer of ice formed when supercooled droplets of cloud or rain freeze on contact with an object.

Gradient: The slope of a surface. It may refer to the ground or an invisible feature such as a pressure surface.

Gradient Wind: A modification of the geostrophic wind which takes centrifugal force into account when the isobars are curved. With cyclonic curvature, the gradient wind is less than the geostrophic wind; with anticyclonic curvature, it is greater.

Graupel: A term used for ice pellets or small hail (diameter less than 5 mm).

Gravity Current: Also called *Density Current* or *Buoyancy Current*. A mainly horizontal current caused by gravity acting on volumes of air or fluid with different densities. Examples are sea-breezes, thunderstorm outflows and bores on rivers and estuaries.

Gravity Wave: Waves caused by gravity and the buoyancy of the air. Lee waves are a type of gravity wave.

Grid: A network, often consisting of squares, drawn on a chart to enable the position of geographic features to be identified by a set of numbers.

Grid Length: The distance between grid points. Large squares are said to have a coarse mesh while small squares have a fine mesh.

Ground Frost: A temperature below 0°C measured an inch or two above the ground. On clear nights, this is several degrees colder than the air temperature, which is measured in a screen at a height of 4 feet.

Gust: A sudden and brief increase in the wind speed, often caused when the flow becomes turbulent over rough ground. A prolonged increase of wind is called a squall.

Haar: Local name for sea fog in eastern parts of Scotland and England.

Haboob: A dust storm in the Sudan caused by a distant thunderstorm squall.

Hail: Small pellets or balls of ice with diameters between 5 and 50 mm (or even more) formed in Cb when raindrops are carried above the freezing level. Large hailstones consist of several layers of ice deposited during successive ascents above the freezing level. If the particles are smaller than 5 mm they are called *Ice Pellets* or *Graupel*.

Halo: The ring of light round the sun or moon caused by refraction and reflection of light by ice crystals, typically in a layer of cirrostratus. The commonest form is a circle of radius 22 degrees.

Haze: Reduction of visibility caused by microscopic dry particles suspended in the atmosphere, usually below an inversion. Some particles are hygroscopic and grow by taking up moisture; this makes the visibility worse.

Heat Low: A depression formed over land which has been heated during the day. It is commonly associated with sea-breezes round the coast and may precede the development of thundery showers.

Helm Wind: Lee wave phenomenon produced when strong easterly winds descend the western slopes of the Crossfell range in Cumbria.

Hertz (Hz): Unit of frequency of one cycle per second.

High: Commonly used to refer to an anticyclone.

Hill Fog: Occurs when high ground becomes covered by low cloud.

Hoar Frost: A deposit of ice crystals in the form of scales, needles, feathers or fans.

Humidity: *Relative Humidity* is the ratio of the actual vapour pressure to the saturated vapour pressure, expressed as a percentage. *Specific Humidity* is the ratio of the mass of water vapour to the total mass of moist air. It is almost the same as the *Humidity Mixing Ratio* which is the ratio of the mass of water vapour to the mass of dry air. This is usually given in grammes of water vapour per kg of dry air.

Humilis (Hum): Applied to small or shallow fair-weather cumulus.

Hurricane: An intense tropical cyclone with winds of 64 knots or more. This name is used in the West Indies. In the western Pacific it is called a *Typhoon.*

Hydrostatic Approximation: This has the effect of filtering out vertically propagating sound waves (not wanted in most Met. equations).

Hydrostatic Equation: Expresses the variation of pressure with height. It is a very good approximation when the air is still or moving horizontally.

Hygrograph: A recording *Hygrometer.*

Hygrometer: An instrument for measuring humidity.

Hygroscopic: Of a substance which absorbs water.

Hypoxia: Lack of oxygen.

ICAO: Abbreviation for *International Civil Aviation Authority.*

Ice: The solid formed when water is cooled beyond its freezing point.

Ice Pellets: Transparent pellets of ice smaller than 5 mm in diameter. They can fall from layered cloud.

Icing (Also called *Ice Accretion*): The formation of ice on an object. Aircraft icing usually occurs when supercooled droplets of water collide with the leading edge. The risk is greatest when the temperature is between 0 and –12°C. Icing can also occur where the pressure falls during the passage of air through engine intakes even though the outside temperature is above zero. The fall of pressure cools the air.

Incus: The Latin for anvil, used to describe the anvil cloud from the top of a Cb.

Initialisation: A process used to smooth out irregularities in the data before starting a numerical forecast.

Insolation: A term to denote the intensity or amount of solar radiation.

Instability Line: A belt of heavy Cu or Cb (not associated with a front) producing showers or thunderstorms.

Intensification: Deepening of a low or strengthening of a high resulting in an increase of wind.

Inversion: A layer of air where the temperature increases upward.

Ion: An electrically charged atom or molecule.

Ionisation: The process of ion formation. Ions render a gas electrically conducting.

Ionosphere: Refers to the atmosphere above about 60 km where the concentration of ions and free electrons reflects radio waves.

IR: Abbreviation for *Infrared Radiation* in the approximate range from 0.7 to 1,000 microns. Satellites use wavelengths between about 10.5 and 12.5 microns for their IR images.

Iridescence: Coloured patches seen on high cloud generally within 30 degrees of the sun caused by diffraction of sunlight by very small cloud particles.

Isallobar: A line of equal barometric tendency. This is usually shown on charts as the change of pressure over 3 hours.

Isallobaric Wind: A component of the *Ageostrophic* wind due to the changing pressure gradient. Geostrophic winds blow parallel to isobars, but isallobaric winds blow at right angles to the isallobars.

Isentropic: A process which does not change the entropy. (Equivalent to an *Adiabatic* process.)

Iso: Equal.

Isobar: A line of constant pressure at a fixed height, usually drawn for MSL on most charts.

Isobaric Surface: A surface of constant pressure.

Isochrones: A series of lines drawn to show the successive positions of a moving system such as a front at specified time intervals.

Isogon: A line along which wind directions are the same.

Isopleth: Any kind of 'iso-' line.

Isotach: A line of constant wind speed.

Isotherm: A line of constant temperature.

Isothermal: Used to describe a layer where the temperature is the same from top to bottom.

Jet Streak: A narrow band of particularly strong winds found within a jet stream, sometimes marked by a detached streak of cirrus. Can occur with an intense and rapidly deepening depression.

Jet Stream: A long belt of very strong winds usually found at high altitudes (mainly between 20,000 and 40,000 feet) caused by a marked horizontal temperature gradient. Often found on the cold side of a frontal system. There are also low-level jets which are smaller and shorter; these influence the development of thundery outbreaks.

Katabatic Wind: A flow of cold (denser) air down a slope, mostly found at night when nocturnal radiation cools the ground and the air in contact with it. Cooling makes the air denser so that it flows down slopes and along valleys.

Katafront: A front where the warm air sinks down above the frontal surface (instead of rising up it). It makes the front weaken and slowly disperses upper cloud.

Kelvin–Helmholtz Waves: Produced when there is strong vertical shear through a shallow atmospheric layer where there is a sharp gradient of temperature and density as at an inversion. The waves gain energy from the large-scale flow.

Kelvin Scale: An absolute temperature scale based on Celsius but with zero equal to −273.15°C.

Khamsin: A hot, dry, southerly wind blowing across Egypt and the Red Sea ahead of a Mediterranean depression. Most frequent between April and June. Khamsin is from the Arabic for fifty, implying fifty days when it is expected.

Knot: A speed of one nautical mile per hour = 1.1508 mph or 0.5144 m/sec.

Labile: A lapse rate equal to or exceeding the *Dry Adiabatic*; used to denote instability.

Land-Breeze: A local wind, usually occurring during the night and early morning, when the cooler, denser air from the land moves out to the warmer sea. It is the reverse of a *Sea-Breeze*.

Lapse Rate: The rate at which temperature decreases with increasing height. The standard atmosphere assumes a lapse rate of 6.5°C per km up to the base of the stratosphere. The *Dry Adiabatic Lapse Rate* is 9.8°C/km.

Latent Heat: The heat absorbed or emitted without change of temperature during a change of state. Heat is given out when water vapour condenses to form droplets and also when water freezes to form ice. Melting and evaporation absorb latent heat.

Layer Clouds: Clouds without vertical protuberances, such as *Stratus*, *Altostratus* or *Cirrostratus*.

Lee Depression: A non-frontal depression formed to the lee of a mountain barrier which is high enough to obstruct the airflow. It is also called an *Orographic Depression*.

Leeward: The region downwind.

Lee Waves: Standing waves which can form when the airflow crosses a range of hills. Air ascends on the upwind side and descends on the downwind side of each wave.

Lenticular: A lens-shaped cloud usually associated with lee waves.

Lidar: Stands for *Light Detection and Ranging*. A system of pulsed light produced by a laser. It is used like radar to study clouds and dust particles in the atmosphere.

Lightning: An electrical discharge from a Cb which causes intense ionisation and a brilliant flash of light often accompanied by a crack of thunder.

Linear System: Equations where alterations in the initial state result in proportional alterations in any subsequent state. Compare with non-linear systems, where alterations in the initial state need not produce proportional changes in a subsequent state.

Line Squall: A long line where there is a rapid increase of wind, usually associated with cold fronts or severe thunderstorms. It may be marked by an arch or line of low black cloud, a rapid rise of wind speed and veer of wind, a rapid drop in temperature and rise in pressure.

Low: Another term for a depression.

Mach Number: The ratio of the air speed to the speed of sound at the corresponding temperature. The speed of sound in the *ICAO* standard atmosphere is about 663 knots at sea level and 573 knots at 11 km (about 36,000 feet).

Mackerel Sky: A sky covered with thin, high altocumulus or cirrocumulus when the cloud has a regular pattern like the scales of a mackerel.

Mamma (also written *Mammatus*): Bulges hanging from the underside of a cloud like udders.

Mares' Tails: A name for streaks of fine cirrus like a horse's tail.

Melting Band: A particularly bright part in the vertical cross-section of a cloud displayed by radar. It occurs where snowflakes melt, giving a stronger return.

Meridional Flow: Airflow along the meridians of longitude, northerly or southerly winds.

Mesh: The spaces in a net. For numerical forecasts, the globe is covered with a geographical grid; the size of the mesh affects the size of weather features which can be analysed.

Mesoscale: Features in the middle scale lying between major lows and highs which dominate a weather map, and local showers which are too small to be shown individually.

Mesosphere: The part of the atmosphere between the *Stratopause* at about 50 km and the *Mesopause* at about 85 km in which the temperature falls with height.

Meteor: The common meaning is a shooting star, a fragment of material which enters the atmosphere from space and burns up with frictional heating. In weather terms it is anything (other than clouds) observed in the atmosphere.

Meteorite: A meteor large enough to reach the earth's surface before burning out.

Meteorology: The science of the atmosphere.

Micro: Used for something very small, usually the millionth part of a unit.

Microbarograph: A barograph designed to record very small and rapid changes of pressure.

Microburst: A small downburst (less than 4 km in diameter), where air descends very rapidly from a Cb causing damage where the air hits the ground and making aircraft landings and take-offs very hazardous.

Micron: Also known as a micrometre, the millionth part of a metre.

Millibar (mbar): One thousandth of a bar. Equals one hectopascal (hPa), equivalent to 1,000 dynes per square cm or 100 newtons per square metre.

Mirage: An optical phenomenon caused by large changes of temperature with height. This alters the refraction of light. Air over a hot desert refracts an image of the sky producing an illusion of water; very cold air makes an object below the normal horizon appear above the surface.

Mist: Reduction of visibility by water drops in suspension. In the UK the term is used when the visibility is 1 km or more and the humidity is over 95 per cent. In the USA this is called *Light Fog*.

Mixing Ratio: The ratio of one gas to the total mass of gases in the atmosphere.

Mizzle: A mixture of mist and drizzle, also called *Scotch Mist*.

Mock Sun (*Parhelion*): An image of the sun caused when light is refracted by ice crystals in cirrus cloud.

Moisture Content (*Specific Humidity*): The ratio of the mass of water vapour to the mass of moist air. Although not identical to the *Mixing Ratio*, the two values are almost the same.

Monsoon: A seasonal pattern of tropical winds around SE Asia, especially the Arabian Sea; in India the SW monsoon is the rainy season while the NE monsoon is dry. By itself, the word monsoon generally refers to a very wet spell.

Mother-of-Pearl Clouds: (also termed *Nacreous Cloud*). An irridescent stratospheric cloud at an average height of 24 km (79,000 feet) possibly formed by standing waves at very high levels. It is best seen after sunset when the sun is still illuminating the cloud.

Mountain Waves: Another term for *Lee Waves*.

MSL: *Mean Sea Level*. The average level of the sea, used as a datum for heights and pressures. In the UK it is based on measurements at Newlyn.

Nacreous Cloud: See *Mother-of-Pearl Cloud*.

Nebulosus: A layer of cloud (such as stratus) with no clear markings.

Nephoscope: A device for measuring the direction and angular velocity of clouds.

Neutercane: A term invented in the USA to describe subtropical cyclones with a radius of much less than 100 nm giving hurricane force winds. The thermal structure of its core is warmer than a sub-tropical cyclone and cooler than a hurricane.

Newton: The force that accelerates a mass of 1 kg by 1 metre per second per second.

Nimbostratus (Ns): A combination of nimbus (rain) and stratus (flattened layer). A deep layer of thick, grey, formless rain cloud.

Noctilucent Cloud: Thin but sometimes brilliant clouds occasionally seen towards the north in summer from latitudes of 50 degrees or more when the sun is not far below the horizon. These clouds are at heights of 80–85 km and seem to move at 100–300 knots.

Normand's Theorem: On an aerological diagram, the dry adiabatic through the dry bulb temperature, the saturated adiabatic through the wet bulb temperature and the saturated mixing ratio line through the dew point all meet at a point. From this, one may derive an approximate condensation level by taking the difference between the surface temperature and dew point (in °C) and multiplying by 400 to get a height in feet, e.g. temp. 20°C, dew point 10°C, cloud base approx. 4,000 feet.

Numerical Weather Prediction: The latest weather observations are converted into numbers which are then processed by computer which solves the equations of motion and physical equations of state. The numerical answers are then converted back into charts, graphs, tables and typed forecasts.

Objective Analysis: The analysis of observations using a computer following mathematical rules. It differs from the subjective analysis made by a meteorologist which may incorporate greater detail but is liable to personal bias.

Objective Forecasts: These are based on a combination of rules and equations, some of them empirical, which are intended to exclude personal bias.

Occlusion: The shutting-off (occluding) of warm air from the surface when a cold front overtakes the preceding warm front. The line where this takes place is called an occlusion.

Okta: Reports of cloud cover are generally given in eighths (oktas) of the sky covered.

Opacus (op): Cloud thick enough to hide the sun.

Orographic Cloud: Cloud formed when the wind blows up a slope. Condensation occurs when the air is cooled below its dew point by the fall of pressure during ascent.

Orographic Low: See *Lee Depression*.

Orographic Rain: The initiation or increase in rainfall due to the cooling of air when it is forced to rise over high ground.

Overturning: Occurs when strong convective currents stir up the air enough to bring upper layers down under the lower layers. It also occurs in a rotor cloud.

Ozone Layer: Ozone is the triatomic form of oxygen which forms a layer between heights of 10–50 km with a maximum concentration between 20 and 25 km. It absorbs UV radiation and this causes a temperature rise near the top, chiefly around 50 km. About 90 per cent of the ozone is at or below about 35 km.

Pannus (pan): Ragged low cloud often found low down in wet and windy weather.

Parametrization: This is carried out by developing semi-empirical equations which relate the large-scale effects calculated at grid points to features such as showers which are too small to be calculated explicitly.

Parhelion (also known as *Sun Dog*): See *Mock Sun*.

Partial Pressure: The pressure exerted by one gas in a mixture. This is the pressure remaining if all other gases are removed but the volume is unchanged.

Pascal (Pa): A unit of pressure. One hectopascal (1 hPa) is 100 pascals and equals one millibar. A number of Met. services have replaced mbar by hPa on charts and diagrams.

Period: The time taken to complete an orbit or oscillation and return to the original value. It is the reciprocal of frequency.

Periodic Orbit: This exactly repeats its past behaviour after a fixed interval.

Perlucidus (pe): A sheet of cloud with gaps.

Phase Line: A vertical cross-section of lee wave flow often shows that the position of wave crests and troughs varies with height. Lines drawn through the crests and troughs at every level are phase lines.

Phase Space: An imaginary space with as many dimensions as the number of variables needed to specify a dynamical system. A simultaneous value of these variables specifies a coordinate in phase space.

Pileus (pil): A small lenticular cap of cloud often formed when the rising top of a cumulus lifts moist air above it. The cumulus may subsequently grow through the pileus.

Pilot Balloon: Originally a small hydrogen-filled balloon released to show the direction a full-sized balloon would drift. Pilot balloons can be tracked by theodolite to find the wind velocity aloft.

Polar Air: Air which has spent a long time at high latitudes and has low temperatures and little moisture content. There are subdivisions such as Polar Arctic which travels fairly directly towards lower latitudes, Polar Maritime which follows a track over the sea, Polar Continental which travels over large land masses, and Returning Polar Maritime which makes a long sweep over the sea and swings back to arrive from a southerly direction.

Polar Front: A front dividing tropical and polar air masses.

Polar Low: A small depression, usually without fronts, which forms in polar air at high latitudes, chiefly during the winter. They often bring heavy snowfall.

Polar Vortex: A cyclonic circulation round the polar regions in both hemispheres. In winter it becomes strong enough to produce a westerly Polar Night Jet Stream at high levels. In summer high-level heating changes the flow to an easterly.

Potential Instability: See *Stability*.

Potential Temperature: The temperature air would have if it were moved along a dry adiabatic until the pressure reached 1,000 mbar. For example, if air at 10°C descended 5,000 feet to the 1,000 mbar level, the temperature would rise to 25°C.

Precipitation: A term which includes drizzle, rain, snow and hail, often used when it is uncertain which will reach the ground: e.g. precipitation may fall as rain on low ground but snow on the mountains.

Pressure: The force, exerted equally in all directions, by a fluid or gas upon a unit area. For Met. purposes, it is usually expressed in millibars (mbar) or their equivalent hectopascals (hPa). 1 mbar = 1 hPa = 0.02953 inches of mercury = 100 newtons per square metre. MSL pressure in the standard atmosphere is 1,013.25 mbar.

Pressure Altitude: The altitude shown by an altimeter with its subscale set to 1,013.25 mbar. It differs from the true altitude if the MSL pressure changes or the temperatures aloft differ from those in the *ICAO* standard atmosphere.

Pressure Gradient: Usually refers to the horizontal force which acts on the air; it is a horizontal vector perpendicular to the isobars (generally at MSL).

Prognostic: Another word for forecast, often abbreviated to *Prog*.

Psychrometer: A pair of matched thermometers mounted together, consisting of a dry bulb measuring air temperature and a wet bulb (covered with muslin kept moist by a supply of pure water). From the difference between the two, one may calculate humidity and dew point. Special tables or a humidity slide rule are used.

Pyrocumulus: Cumulus formed above a large fire on the ground.

Q-Code: Used originally for abbreviating wireless telegraphy messages. Some items are still used in plain language, e.g.:

QBA	visibility
QBB	cloud base
QFA	Met. forecast
QFE	station-level pressure
QFF	sea-level pressure
QNE	airfield altitude with altimeter set to 1,013.25 mbar
QNH	altimeter setting (the altimeter then reads the true height of the airfield for landing)
QNT	Gust

Radar: *Radio Detection and Ranging*. Originally designed for locating aircraft, subsequently found useful for drawing a radar map of the terrain and later used for tracking precipitation. (See also *Doppler Radar*.)

Radiation: Transmission of energy by electro-magnetic waves.

Radiation Fog: Fog formed when the land cools by radiating away heat on a clear night.

Radiatus (ra): Term applied chiefly to cirrus and altocumulus clouds which are arranged in bands which seem to radiate from a point on the horizon.

Radio Duct: The result of a change in the refractive index which keeps a radio beam trapped within a sort of duct near the ground. Also referred to as *Anaprop* (*Anomalous Propagation*).

Radiometer: Instrument for measuring the flux of radiation.

Radiosonde: An expendable transmitter carried up by balloon to measure pressure, temperature and humidity.

Radio-theodolite: Used for measuring the bearing and angular elevation of a distant transmitter such as a radiosonde.

Rain: Falling drops of water of diameter greater than 0.5 mm.

Rainbow: Caused by the reflection and refraction of sunlight by drops of water producing a ring of colour making an angle of about 42 degrees or less from the shadow of the observer's head. With double reflection in the raindrops the angle is about 50 degrees, producing a double rainbow. Bows may be seen by moonlight and in fog, but colours cannot be distinguished in darkness.

Rain Shadow: The sheltered (lee) side of mountains, where rainfall is much reduced.

Ravine Wind: Strong winds which funnel down a mountain valley.

Remote Sounding: Satellites can measure the temperature profile in the atmosphere vertically below them by means of an infrared spectrometer.

Revolving Storm: Another term for a tropical cyclone, hurricane or typhoon.

Ridge: Extension from an anticyclone sometimes called a *Wedge* of high pressure.

Rime: Rough, white ice crystals which form when supercooled droplets collide with a freezing object.

Roll Cloud: A long bar of cloud which forms when cold, dense air lifts and flows under a warmer moist layer. It can be caused by thunderstorm outflows, sea-breeze fronts or undular bores.

Rope Cloud: A long line of cloud which looks like a rope on satellite pictures. It forms at a frontal wind shift or squall line.

Rossby Wave: A wave-shaped contour pattern usually found on upper air charts. The wavelength is often 1,000 nm or more. Four or five Rossby waves may extend round a hemisphere, chiefly at middle latitudes.

Rotor: A rotating eddy with horizontal axis found on the lee side of a mountain range when lee waves are present. It occurs under the crest of the primary lee wave, where it can cause very severe turbulence. The surface wind may be reversed below it.

Runway Visual Range (RVR): The maximum distance along the runway centre line at which specified lights or markers can be seen (theoretically from a pilot's eye view about 15 feet above the ground). It may be very different from the general visibility in the area.

Satellite Sounding: Measurement of the temperature profile vertically below a satellite by measuring the radiation emitted by carbon dioxide and water vapour at a number of wavelengths.

Saturated Adiabatic Lapse Rate (SALR): The rate of decrease of temperature when saturated air is lifted. On aerological diagrams, the SALR is shown by a family of curves. These assume that condensed water falls out and is not carried up higher. Unlike a dry adiabatic, the process is not reversible.

Saturation: Moist air is saturated if it is holding the maximum amount of water vapour for the temperature and pressure. If there are not enough condensation nuclei present, the air may become *Supersaturated*.

Scotch Mist: A mixture of mist and drizzle typically found when moist air blows over the hills of Scotland. It is called mizzle in SW England.

Scud: Another name for *Fracto-Stratus*, ragged low cloud usually associated with strong winds and rain.

Sea-Breeze: The flow of air from the sea to districts inland which have been heated by the sun.

Sea-Breeze Front: A convergence line, often marked by cumulus clouds, which forms when the sea-breeze pushes inland against a wind from the land.

Sea Smoke: See *Arctic Sea Smoke*.

Secondary Cold Front: One or more troughs in the cold air following a cold front. Instead of scattered showers, the precipitation is almost continuous along it.

Secondary Depression: A small low which forms within the circulation of an older and deeper depression. It may start as a wave on a trailing cold front or (less often) on the warm front. Secondaries often become much deeper, develop very strong winds and may eventually absorb the original low.

Seeding: The stimulation of rain by introducing ice crystals into the cloud, often by dropping solid carbon dioxide particles into it. Silver iodide, released from the ground as a fine smoke, is also used as an ice-nucleation agent.

Semi-Geostrophic: A modification of the *Geostrophic* wind which incorporates the ageostrophic part of the horizontal wind.

Sensible Heat: Heat which can be felt or measured, as distinct from *Latent Heat* which is locked into water vapour until condensation occurs.

Sferics: A term used for the position of lightning flashes determined by means of radio. (An abbreviation of *Atmospherics*.)

Shade Temperature: Also termed *Screen Temperature*, the value measured by a thermometer screened from direct sunlight and heat radiated from the ground but open to the flow of air.

Shear of Wind: The rate of change of wind velocity with distance (horizontal shear) or height (vertical shear).

Shear-Gravity Wave: This forms when there is wind shear at the boundary of two layers with different density.

Shear Wave: Waves which form within a stable layer which has a strong vertical wind shear across it.

Showalter Index: Used chiefly in the USA for the prediction of thunderstorms.

Shower: Rain, hail or snow from a convective cloud, usually intermittent or fairly brief but often heavier than continuous precipitation.

Sigmet: Short for Significant Meteorological phenomenon, used as a prefix to warning messages.

Significant Weather Chart: A forecast weather map usually only valid for certain flight levels, which shows items specified as being significant for the type of flight. Fronts and areas of bad weather such as thunderstorms, ice-producing clouds and turbulence are generally marked, but areas of fog, low cloud and some types of precipitation may be omitted for higher-level flights.

Sink: (a) the rate of descent of a glider or (b) the point or region where something vanishes or is dissipated. The opposite term is *Source*.

Slant Visibility: The greatest distance at which an unlit object can be identified when looking down at an angle. It may differ from the horizontal visibility reported by an observer on the ground, particularly if there is a shallow layer of mist or fog.

Sleet: No agreed international meaning, but in the UK it means snow melting as it falls, or a mixture of snow and rain.

Sling Psychrometer: A psychrometer mounted on a rotating frame like a football rattle. Whirling it round creates a flow of air over the bulbs.

Slush: Partly melted snow or ice; the drag from deep slush has caused take-off accidents.

Smog: A contraction of *Smoke Fog*, a mixture of water fog and pollutants such as smoke or other chemical particles. The term is also used for the photochemical haze produced when sunlight reacts with exhaust and other fumes.

Snow: Solid precipitation in the form of ice crystals which link together to form snowflakes at temperatures near zero. Granular snow consists of small grains less than 1 mm in diameter.

Solar Constant: Radiation from the sun received outside the earth's atmosphere, approximately 139.6 milliwatts per square cm or 1.4 kW per square metre.

Source: The point or region where something is created and from which it moves outward.

Specific Heat: The temperature needed to raise the temperature of a unit mass by one degree. It is usually given in joules per kg per degree Kelvin or in calories per gramme per degree K.

Specific Humidity: The ratio of the mass of water vapour to the mass of moist air. It is almost the same as the *Humidity Mixing Ratio*.

Squall: A sudden increase of wind by at least 16 knots with the mean speed rising to 22 knots or more and lasting for at least a minute.

Squall Line: A long line along which the wind rises to a squall, usually produced by a Cb or a cold front.

Stability: Static stability can be tested by observing how a parcel of air behaves if displaced from its original level. It is stable if it returns to the original level, neutral if it remains at the new level and unstable if it continues to rise or fall in the direction of displacement. Dynamic stability means there is no tendency for small wave-like disturbances to grow.

Standard Atmosphere: An average state of the atmosphere as specified by *ICAO* and used for calibrating altimeters. It has a lapse rate of 6.5°C from a surface pressure of 1,013.25 mbar and temperature of 15°C up to the base of the stratosphere, assumed to be at 11 km (about 36,000 feet). Above that, the lapse rate is almost isothermal.

Standard Pressure: 1,013.25 mbar or hPa; the pressure exerted by a column of mercury 760 mm high of density 13,595.1 kg per cubic metre subject to gravity of 9.80665 m per second per second.

Standing Wave: A wave which moves upwind at the same speed as the air travels downwind, thus remaining stationary over the ground. Lee waves are usually also standing waves.

Steam Fog: An alternative for *Arctic Smoke Fog*.

Steering: A factor controlling the movement of lows and highs. The pattern of airflow at high levels generally determines the direction and speed of motion.

Stevenson Screen: A standard housing for thermometers. It is painted white and the sides have double louvres to give good airflow.

Storm: Storm-force winds have a mean speed greater than 47 knots (Beaufort Force 10). The term applies to various kinds of violent weather.

Stratiformis (str.): Clouds spread out in a horizontal layer.

Stratosphere: A deep, very dry and stable layer lying above the *Troposphere* and separated from it by the *Tropopause*. The base of the stratosphere varies, being highest in tropical regions and lowest in cold polar regions. The stratosphere generally lies between 10 and 50 km and contains the ozone layer.

Stratocumulus (Sc): A layer of cloud with rolls or rounded elements like flattened Cu. It often forms when convective currents carrying moist air from the surface spread out under an inversion.

Stratus (st.): A flat, featureless layer of cloud formed when the air is cooled below its dew point either by passage over a cold surface or by smooth lifting up a frontal surface or hillside.

Streamline: A line parallel to the wind. Unlike isobars, the streamlines can join together or fan out showing convergence or divergence; they are used in equatorial regions, where isobars seldom represent the wind adequately.

Sublimation: A term used for direct evaporation from an ice surface or direct deposition of ice from water vapour.

Subsidence: The slow downward motion of air over a wide area. It occurs most strongly where pressure is rising during the development of a ridge or high. The slow sinking warms and dries the air and often results in the formation of an inversion above the surface.

Subtropical High: A semi-permanent region of high pressure normally found between the tropics and the temperate zones, usually between latitudes 23 and 40 degrees.

Sun Dog: See *Mock Sun*.

Sun Pillar: A vertical column of light above or below the sun due to reflection from ice crystals.

Sunspot: A relatively dark region on the sun. The features are quasi-periodic with a maximum every eleven years. They affect the ionosphere and were once thought to influence the rainfall.

Superadiabatic Lapse Rate: This is greater than the adiabatic lapse rate of 9.8°C per km. It normally only exists near the surface when sunshine heats the ground strongly.

Supercell: A single-celled Cb formed when there is vertical wind shear. Ordinary Cb have several cells with surges of convective updrafts, but the supercell has a continuous circulation with air entering at low level and leaving at the top where it is blown away by the strong winds to form an anvil. Supercells have a long life; the whole cloud may rotate and set off tornadoes as well as severe hail and thunderstorms.

Supercooling: Liquid cooled below its nominal freezing point. Most cloud droplets can remain liquid well below 0°C, while some remain liquid down to −40°C. Supercooled drops freeze on contact to produce airframe icing.

Supersaturation: Air containing more than enough water vapour to produce saturation.

Surface Inversion: Exists when the temperature increases above the surface. It is formed either by the ground being cooled by radiation on a clear night or because warmer air is advected across a cold surface.

Surface Temperature: The term usually refers to the screen temperature 4 feet above the ground. Ground temperature is measured by a thermometer on the ground. For the grass temperature, the thermometer bulb just touches the tips of short grass.

Surface Wind: This refers to the wind at a height of 10 m in an unobstructed area.

Synoptic Chart: A term for a weather map drawn for a fixed time.

Synoptic Meteorology: The study of current weather conditions in order to predict future developments.

Synoptic Station: A place where weather observations are made at fixed times for transmission to a centre where synoptic charts are plotted.

Teleconnection: A linkage between different Met. features separated by great distances. For example, statistics show that the rainfall in the southern Sahara (Sahel) varies with the phase of the El Niño Southern Oscillation (ENSO) in the Pacific.

Temperature Scales: (a) *Celsius* (C), previously called Centigrade, has the freezing point of water at 0°C and the boiling point at 100°C; (b) *Fahrenheit* (*F*) has the freezing point at 32°F and the boiling point at 212°F; (c) *Kelvin* (*K*), also known as the absolute temperature, has zero at the lowest possible temperature (−273.15°C) and boiling point at 373.15°C. Conversion formulae:

$$F = (C \times \tfrac{9}{5}) + 32$$
$$C = (F − 32) \times \tfrac{5}{9}$$
$$K = C + 273.15$$

Tendency (in Met.): The rate of change of an element, mostly used for the change of barometric pressure over a period of three hours.

Tephigram: An aerological diagram with the x, y coordinates temperature (T) and Entropy (phi). The lines of equal entropy are also the dry adiabatics. It is used for plotting temperature and humidity at specified pressure levels. Stability can be worked out by comparing the plotted lapse rate with the lines representing dry and wet adiabatics.

Thermal: A volume of air which rises because it is warmer (less dense) than its environment. Many thermals form when the air near the surface is warmed by contact with ground heated by the sun.

Thermal Wind: The geostrophic vertical wind shear between the base and top of a deep layer, often chosen to be from 1,000 to 500 mbar. It depends on the horizontal temperature gradient; in the northern hemisphere, the thermal wind blows so that low temperatures lie on the left-hand side.

Thermodynamic Diagram: Another name for an aerological diagram.

Thermodynamics: This deals with the transfer of heat into energy and energy into heat.

Thermograph: A recording thermometer.

Thermometer Screen: A housing which shields the instruments from sunlight or radiation from the ground but allows air to pass through. The thermometer bulbs are generally 4 feet above the ground.

Thickness: The term used for the depth of a layer of air (often the layer between 1,000 and 500 mbar). The thickness varies with the temperature: warm air expands so its thickness is greater than for a cold column.

Thunder: The noise caused by the shock wave from a narrow channel of air heated by lightning to around 28,000°C.

Thunderstorm: Rain, hail or snow shower accompanied by flashes of lightning from a Cb cloud.

Tornado: A destructive vortex formed by extremely rapid rotation of air in and below the base of a Cb. The rate of spin reduces air pressure and causes condensation to form a funnel-shaped cloud extending down to the ground. Rotational speed may reach 300 knots, enough to destroy buildings along its track which may be less than 100 yards wide but can extend to a mile. Wide tornadoes sometimes contain several smaller 'suction vortices' rotating round the centre.

Torro Index: A scale of tornado intensity.

Trade Winds: Winds with a predominantly easterly component which spread out from the subtropical anticyclones towards the equator.

Trajectory: The path followed by an object (in Met. usually a mass of air, e.g. the path of a depression).

Translucidus (tr.): Thin layer clouds through which the sun can be seen.

Tropical Air: A very warm air mass originating in subtropical anticyclones near latitude 30 degrees. It can be very moist after a long sea passage (*Tropical Maritime*). If it comes mostly overland it is *Tropical Continental* and gives very high temperatures in summer.

Tropical Cyclone: A vigorous depression in tropical latitudes producing winds of gale or hurricane force. Similar systems are called *Hurricanes* in the West Indies and the USA but *Typhoons* in the China Sea and western Pacific.

Tropopause: The atmospheric boundary between the troposphere and the stratosphere. It is usually marked by a kink in the temperature profile where the lapse rate changes from about 6.5°C per km to almost isothermal.

Troposphere: The lower layer of the atmosphere where most of the weather takes place. On average, the temperature decreases with height up to the tropopause.

Trough (of low pressure): Appears on weather maps as an extension from a depression.

Trowal: A Canadian term for a trough with warm air aloft. (Rather similar to an *Occlusion*.)

Tuba (tub.): Also known as *Funnel Cloud*, and is a column or inverted cone of cloud extending below the base of a Cb indicating a vortex. It may be the first indication of a tornado or waterspout.

Turbulence: Disturbed air motions caused by numerous eddies of different size, making conditions rough for flying.

Typhoon: The name for a severe tropical storm in the China Sea or western Pacific.

Upper Cold Front: The leading edge of dry air which may move ahead of the surface cold front and overrun the warm conveyor belt. Sometimes called a 'split cold front'.

Upper Trough: A trough in the contours of a pressure surface high above the ground but not necessarily present at the surface. Associated with cold unstable air aloft, and frequently follows a cold front.

Upslope Fog: Fog which forms when moist air flows up the windward side of high ground. The air is cooled below its dew point by adiabatic expansion during lifting.

UT (*Universal Time*): Defined by the rotation of the earth, it used to be called **GMT (*Greenwich Mean Time*)** and is distributed by broadcast time signals.

UV (*Ultraviolet*): Light whose wavelength is shorter than the visible spectrum, often associated with sunburn. It is strongly absorbed by **Ozone** in the stratosphere but in recent years has penetrated to low levels over polar regions, where an **Ozone Hole** developed during the winter months.

Valley Wind: A wind blowing up the valley when the sun heats the mountain slopes above; a type of **Anabatic Wind.**

Vapour: A gas which is below its 'critical temperature', that is, when it can be liquefied by pressure alone.

Vapour Pressure: The fraction of the atmospheric pressure due to the presence of water vapour. It can be measured indirectly from dry- and wet-bulb readings using a humidity slide rule or psychrometric tables.

Vector: A quantity defined by both direction and magnitude, e.g. a wind arrow which shows both the direction and speed of the wind.

Veering: A change of wind in a clockwise direction. The reverse is termed 'backing'; e.g. the wind veers from north to north-east but backs from north to north-west.

Velocity: A **Vector** showing the rate of change of position in a specified direction.

Virga (vir.): Also known as *Fallstreaks*; trails of precipitation from a cloud. They evaporate before reaching the ground.

Virtual Temperature: The temperature at which dry air would have the same density as moist air. The introduction of moisture makes the air less dense. The difference between virtual and actual temperature is very small in cold air which cannot hold much moisture, but in tropical air the virtual temperature may be several degrees higher than the actual temperature.

Viscosity: The property of a fluid which resists deformation, e.g. oil is more viscous than water.

Visibility: (For Met. purposes) the greatest distance a dark object can be seen against the sky by daytime. (At night it may need to be established by reference to lights of known intensity.)

Vortex: A rotating swirl of air or water (such as the conical depression where the water goes down the plug hole of a bath). The axis of rotation is a **Vortex Line.** The term vortex is also used for large, almost circular, non-frontal depressions.

Vortex Shedding: A term used to describe airflow past a vertical obstruction such as a cylindrical chimney or a mountainous island. A series of vortices breaks away from the lee side; they rotate alternately in opposite directions and may be marked by long, wandering trails of cloud extending from peaks on isolated islands.

Vorticity: The spin of a fluid defined as twice the rate of rotation. In the atmosphere, vorticity is calculated from the curvature of the flow and the wind shear. Cyclonic vorticity is given a positive sign, anticyclonic is negative. Relative vorticity is calculated relative to the rotating earth. Absolute vorticity is related to coordinates in space; it is determined by adding the rotation of the earth to the relative vorticity. Vorticity is increased where air converges and decreased when it spreads out (diverges).

Warm Conveyor Belt: A band of warm moist air drawn into the flow round a trough or low; it rises above a frontal surface and forms cloud and rain along and ahead of the front.

Warm Front: A front which moves so that cold air is replaced by warm air.

Warm-Front Wave: A small depression which starts as a wave on the warm front and moves away from the parent low.

Warm Sector: The sector of a depression, usually triangular in shape, which contains the warmest air. It is bounded by warm and cold fronts and grows narrower as the cold front overtakes the warm during the process of *Occlusion*.

Waterspout: A very rapidly rotating funnel-shaped cloud extending down from the base of a Cb to the sea, where it creates a disturbance. It is the marine equivalent of a tornado.

Water Vapour: Water in the form of an invisible vapour; the gas formed when water is evaporated.

Wave Clouds: Clouds which form where air rises in the crests of waves, often but not invariably associated with airflow across mountains.

Wave Depression: A depression which forms at the tip of a wave-like undulation on a front.

Wavelength: The distance between successive wave crests.

Wave Motion: Oscillatory movement in any medium which results in waves moving through it.

Wave Number: Either the number of waves in unit distance (the reciprocal of the wavelength) or 2 pi times this number.

Wedge of High Pressure: Another term for a ridge.

Wendy Windblows: The punning name given to a network of automatic weather stations on hilltops in England and Wales sponsored by hang-glider pilots. Subscribers are given code numbers to interrogate the reports by telephone.

Wet Adiabatic: Another term for the *Saturated Adiabatic Lapse Rate*.

Wet-Bulb: A thermometer with wet muslin over the bulb. Evaporation cools the thermometer.

Wet-Bulb Depression: The difference between the wet- and dry-bulb temperatures. The drier the air, the greater the difference. *Humidity* and *Dew Point* can be calculated from these values.

Whirlwind: A small, rapidly rotating column of air such as a *Dust Devil*.

White-out: Descriptive of conditions over a snow field when a cloud sheet eliminates all shadows. There is no visual contrast to distinguish ground features.

Wind: The movement of air, generally horizontal, usually reported in knots for aviation but in metres per second for scientific papers.
1 knot = 0.51444 m/sec = 1.852 km/h = 1.1508 mph.

Wind Shear: The rate of change of the vector wind with distance at right angles to the wind direction.

Wind Vane: Usually a horizontal arm mounted on a pivot and stabilised by a fin at one end which keeps it pointing into the wind.

Windward: The direction into wind. The opposite of *Leeward*.

WMO: Abbreviation for the *World Meteorological Organisation* at Geneva.

Zonal Flow: Airflow parallel to lines of latitude, usually westerly winds. Easterlies are considered as negative zonal flow.

Appendix 1

Abbreviations used in Met. documentation

AAL	Above airfield level	CLD	Cloud
AC	Altocumulus	COR	Correction (sometimes CCA, CCB, etc.)
ACT	Active or activated	COT	At the coast
ADJ	Adjacent	CNS	Continuous
AGL	Above ground level	CS	Cirrostratus
AIREP	Air report	CU	Cumulus
AMD	Amended or amend	CUF	Cumuliform
AMSL	Above mean sea level	CW	Clockwise
ARMET	Area Met. (forecast of upper winds and temperatures)	DALR	Dry adiabatic lapse rate
AS	Altostratus	DECR	Decrease
ASSW	Associated with	DEG	Degrees
ASR	Altimeter setting region	DENEB	Fog dispersal operations
ATC	Air Traffic Control	DIF	Diffuse
ATCC	Air Traffic Control Centre	DP	Dew point
		DR	Low drifting
BASE	Cloud base	DRSN	Drifting snow
BC	Patches	DTRT	Deteriorate or deteriorating
BCFG	Fog patches	DUC	Dense upper cloud
BKN	Broken (5/8–7/8 cloud)	DZ	Drizzle
BL	Blowing		
BLO	Below clouds	ELR	Environmental lapse rate
BLSN	Blowing snow	EMBD	Embedded (in cloud) e.g. CB EMBD in AS
BR	Mist (from French word 'brume')		
BTL	Between layers	EXP	Expect or expected
BTN	Between	EXTD	Extend
C	Degrees Celsius (centigrade)	F	Degrees Fahrenheit
CAST	Castellanus (turret cloud)	FA	Area forecast
CAT	Clear air turbulence	FAX	Facsimile
CAVOK	Ceiling and visibility OK (and no 'weather')	FBL	Feeble, light (of icing, etc.)
		FC	Funnel cloud
CB	Cumulonimbus	FCST	Forecast
CBR	Cloud base recorder	FG	Fog
CC	Cirrocumulus or counter clockwise	FIC	Flight Information Centre
CI	Cirrus	FIS	Flight Information Service
CFO	Central Forecast Office	FIR	Flight Information Region
CIT	City (near or over large towns)	FLUC	Fluctuating
CLA	Clear icing	FM	From

FOQNH	Forecast QNH		**LOC**	Locally
FPM	Feet per minute		**LSQ**	Line squall
FR	Route forecast		**LTD**	Limited
FRONT	Weather front		**LV**	Light and variable
FRQ	Frequent		**LYR**	Layer or layered
FT	Feet			
FU	Smoke (from French 'fumée') also forecast upper winds		**M**	Metres (or minus if put before a temperature)
FZ	Freezing		**MAR**	Sea (at or over the sea)
FZDZ	Freezing drizzle		**MAX**	Maximum
FZFG	Freezing fog		**MB**	Millibar
FZRA	Freezing rain		**METAR**	Met. Airfield Report
FWA	Flight Watch Area		**MET(O)**	Met. (Office)
FWC	Flight Watch Centre		**MI**	Shallow
			MIFG	Shallow fog
GND	Ground		**MNM**	Minimum
GR	Hail (from German 'graupel', soft hail)		**MOD**	Moderate
GRADU	Gradually		**MON**	Above mountains
GST	Gust		**MOTNE**	Met. Operational Telecom Network (Europe)
HGT	Height		**MOV**	Move or moving
HPA	Hectopascals (same as millibars)		**MPS**	Metres per second
HRCN	Hurricane		**MS**	Minus
HZ	Haze		**MSL**	Mean sea level
			MT	Mountain
IAO	In and out (of cloud)		**MTW**	Mountain wave
ICAO	International Civil Aviation Organisation		**MX**	Mixed clear and rime ice
ICE	Icing		**NC**	No change
IFR	Instrument Flight Rules		**NIL**	None
IMC	Instrument Met. conditions		**NOSIG**	No significant change
IMPR	Improve		**NM**	Nautical mile
IMT	Immediate or immediately		**NS**	Nimbostratus
INC	In cloud			
INCR	Increase		**OBS**	Observed
INS	Inches		**OBSC**	Obscured
INTER	Intermittent		**OCNL**	Occasional
INTSF	Intensify		**OKTA**	1/8 of sky area
INTST	Intensity		**OPA**	Opaque rime ice
ISOL	Isolated		**OTP**	On top
IR	Ice on runway		**OVC**	Overcast (8/8 cloud)
JET	Jet stream		**PE**	Ice pellets
JTST	Jet stream		**PO**	Dust devils
			PROB	Probability
KM	Kilometres		**PROV**	Provisional
KM(H)	Kilometres (per hour)		**PS**	Plus
KT	Knot		**PSN**	Position
LAN	Land (inland)		**QFA**	Meteorological forecast

QFE	Airfield pressure setting	SYRED	Synoptic (reduced length)
QFF	Sea level pressure		
QNE	Altitude on landing with sub-scale set on 1013.2 mbar	T	Temperature
		TAF	Terminal airfield forecast
		TC	Tropical cyclone
QNH	Altimeter setting which gives airfield altitude on landing	TCU	Towering cumulus
		TDO	Tornado
QNT	Gust speed	TEND	Tending to (a weather trend)
		TEMPO	Temporarily
RA	Rain	TIL	Until
RAG	Ragged	TIP	Until past (place)
RAPID	Rapid (used for weather changes)	TOP	Cloud top
RASH	Rain shower	T/P	Teleprinter
RASN	Rain and snow	TRS	Tropical revolving storm
RE	Recent (within last hour but now ceased)	TS	Thunderstorm
		TSGR	Thunderstorm with hail
RTD	Retard (a late report)	TSSA	Thunderstorm with sand/duststorm
RVR	Runway visual range	TURB	Turbulence
RWY	Runway	TYPH	Typhoon
SA	Sandstorm or duststorm		
SALR	Saturated adiabatic lapse rate	UA	Air report
SC	Stratocumulus		
SCT	Scattered (1/8–4/8 cloud)	VAL	In valleys
SEV	Severe	VER	Vertical
SFC	Surface	VERVIS	Vertical visibility
SFLOC	Sferic location. Radio location of lightning flashes	VFR	Visual Flight Rules
		VIS	Visibility
SG	Snow grains	VMC	Visual meteorological conditions
SH	Showers	VOLMET	Voice Met. broadcast (R/T)
SIGMET	Significant Met. (weather warnings)	VRB	Variable
SIGWX	Significant weather	VSP	Vertical speed
SKC	Sky clear		
SLW	Slow	WDSPR	Widespread
SN	Snow	WKN	Weaken
SNSH	Snow showers	WRNG	Warning
SPECI	Special weather report	WS	SIGMET or wind shear
SPOT	Spot winds	WTSPT	Waterspout
SQ	Squall	WX	Weather
ST	Stratus		
STF	Stratiform	XS	Atmospherics
STNRY	Stationary	XX	Heavy (qualifies weather such as showers) e.g. XXTS means heavy thunderstorm
SYNOP	Synoptic. International coded weather report		

LERWICK

WICK

STORNOWAY

ABERDEEN

TIREE

LEUCHARS
FIFENESS

GREENOCK

MALIN HEAD

MACHRIHANISH

BOULMER

LARNE

RONALDSWAY

BRIDLINGTON

LIVERPOOL CROSBY

VALLEY

VALENTIA

ABERPORTH

MILFORD HAVEN

SHEERNESS

SANDETTE LV

ST. CATHERINE'S PT

GREENWICH LV

SCILLY

CHANNEL LV

JERSEY

Coastal reports

Code figure	N	W₁W₂	C_L	C_M	C_H	C	a	D_s
0								
1								
2								
3								
4								
5								
6								
7								
8								
9								
/								

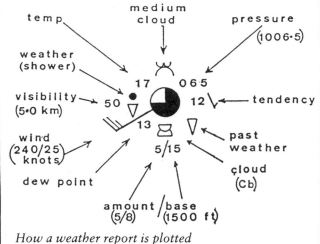

How a weather report is plotted

Table of symbols for

cloud amount	(N)	high cloud	(Ch)
past weather	(W)	general cloud	(C)
low cloud	(Cl)	pressure tendency	(A)
medium cloud	(Cm)	ships' course	(Ds)

Table of symbols for present weather

ww	0	1	2	3	4	5	6	7	8	9
0										
1										
2										
3										
4										
5										
6										
7										
8										
9										

Present weather code (shortened version) see table of symbols opposite

04	Smoke	05	Haze
06	Dust in suspension	07	Rising dust
08	Dust devil	09	Adjacent sandstorm
10	Mist (visibility more than 1 km)	11	Patches of shallow fog
12	Continuous shallow fog	13	Lightning seen
14	PPN in sight (not reaching ground)	15	PPN in sight (more than 5 km away)
16	PPN in sight (less than 5 km away)	17	Adjacent thunderstorm
18	Squalls	19	Funnel clouds in past hour
20	Recent drizzle (in last hour)	21	Recent rain (in last hour)
22	Recent snow (in last hour)	23	Recent sleet (in last hour)
24	Recent freezing rain	25	Recent rain shower
26	Recent snow shower	27	Recent hail shower
28	Recent fog (in last hour)	29	Recent thunderstorms
30	Dust/sandstorm (decreasing)	31	Dust/sandstorm
32	Dust/sandstorm (increasing)	33	Heavy dust/sandstorm (decreasing)
34	Heavy dust/sandstorm	35	Heavy dust/sandstorm (increasing)
36	Low drifting snow (mod)	37	Low drifting snow (severe)
38	Blowing snow (mod)	39	Blowing snow (severe)
40	Adjacent fog or fog banks	41	Fog patches affecting station
42	Fog thinning (sky visible)	43	Fog thinning (sky obscured)
44	Fog, no change (sky visible)	45	Fog, no change (sky obscured)
46	Fog thickening (sky visible)	47	Fog thickening (sky obscured)
48	Freezing fog (sky visible)	49	Freezing fog (sky obscured)
50	Intermittent slight drizzle	51	Continuous slight drizzle
52	Intermittent moderate drizzle	53	Continuous moderate drizzle
54	Intermittent heavy drizzle	55	Continuous heavy drizzle
56	Freezing drizzle (slight)	57	Freezing drizzle (mod to heavy)
58	Drizzle and rain (slight)	59	Drizzle and rain (heavy)
60	Intermittent slight rain	61	Continuous slight rain
62	Intermittent moderate rain	63	Continuous moderate rain
64	Intermittent heavy rain	65	Continuous heavy rain
66	Freezing rain (slight)	67	Freezing rain (mod/heavy)
68	Rain and snow (slight)	69	Rain and snow (mod/heavy)
70	Intermittent slight snow	71	Continuous slight snow
72	Intermittent moderate snow	73	Continuous moderate snow
74	Intermittent heavy snow	75	Continuous heavy snow
76	Ice prisms	77	Snow grains
78	Snow crystals	79	Ice pellets
80	Showers of rain (slight)	81	Showers of rain (mod/heavy)
82	Showers of rain (violent)	83	Showers of rain and snow (slight)
84	Showers of rain and snow (mod/heavy)	85	Showers of snow (slight)
86	Showers of snow (mod/heavy)	87	Showers of soft hail (slight)
88	Showers of soft hail (mod/heavy)	89	Showers of hail (slight)
90	Showers of hail (mod/heavy)	91	Slight rain after thunderstorm
92	Mod/heavy rain after thunderstorm	93	Slight snow or hail after thunderstorm
94	Mod/heavy snow or hail after thunder	95	Thunderstorm (sl/mod rain or snow)
96	Thunderstorm (sl/mod hail)	97	Thunderstorm with heavy rain or snow
98	Thunderstorm with heavy sand/dust	99	Thunderstorm with heavy hail

Appendix 2

Weather in METAR Reports

A METAR is an aerodrome actual weather report. Messages appear in varying degrees of complexity. Major airports such as London Heathrow (EGLL) add extra details such as variations of wind and runway visual range, but minor airports usually send simple messages, such as the example below:

EGPA   29023KT   7000   DZ   SCT003   BKN006   OVC009   07/06   Q1016   TEMPO   9999 BKN010=

Meaning

EGPA	Airfield:	Kirkwall
29023KT	Wind:	290 degrees 23 knots
7000	Visibility:	7000 metres (7 km)
DZ	Weather:	drizzle
SCT003	Lowest cloud:	scattered at 300 feet
BKN006	Second cloud layer:	broken at 600 feet
OVC009	Third cloud layer:	overcast at 900 feet
07/06	Temp & dew point:	temperature 7, dew point 6
Q1016	QNH pressure:	QNH 1016 millibars
TEMPO	Temporary variation	
9999	Visibility:	10 kilometres or more
BKN010	Cloud:	broken at 1000 feet

Gusts may be reported when the maximum wind is 10 knots or more above the mean wind reported, e.g.

EGPA 29023G35KT . . .

Meaning

EGPA	Airfield:	Kirkwall
29023G35KT	Wind:	290 degrees 23 knots max 35 knots

Major airports, such as London Heathrow, add variations in the wind direction and (in poor visibility) the runway visual range, e.g.

EGLL   24005KT   200V280   R27R/110 . . .

Meaning

EGLL	Airfield:	London Heathrow
24005	Wind:	240 degrees 05 knots
200V280	Variation:	varying between 200 and 280 degrees
R27R/110	Runway Visual Range:	runway 27 right 1100 metres

Weather abbreviation in METAR and TAF messages

BC	Patches	FZ	Freezing	SH	Shower
BR	Mist	GR/GS	Hail	SS	Sandstorm
BL	Blowing	HZ	Haze	SN	Snow
DR	Drifting	MI	Shallow	SQ	Squalls
DU	Dust	PE	Ice pellets	TS	Thunderstorm
DZ	Drizzle	RA	Rain	VC	In vicinity
FG	Fog	RE	Recent	WS	Wind shear
FU	Smoke	SA	Sand		

Wind shear is not at present reported by UK airfields, but may be reported in the USA, e.g. (a) WS TKOF RWY24 means a wind shear on take-off path from Runway 24; (b) WS LDG 27 means wind shear on approach path to Runway 27.

Intensity is shown by adding + or − before the letters, e.g. −DZ means light drizzle, +RA means heavy rain. No + or − indicates 'moderate', e.g.
−SN means slight snow, SN moderate snow, +SN heavy snow.

Abbreviations can be combined, e.g. RETS means recent thunderstorm, FZDZ freezing drizzle, MIFG shallow fog.

Cloud group abbreviations

BKN	Broken (5/8 to 7/8)	SCT	Scattered (3/8 to 4/8)
FEW	1/8 to 7/8	SKC	Sky clear
OVC	Overcast (8/8 cover)	VV///	Sky obscured
CAVOK	Ceiling and visibility OK		

Note

CAVOK is reported in the following conditions: if visibility is more than 10 km; no cumulonimbus; no cloud below 5000 ft or highest minimum sector altitude; no precipitation, shallow fog or low drifting snow.

Cloud types are only reported if there are:

CB	cumulonimbus	TCU	Towering cumulus

Trend abbreviations

A trend is an optional addition which comes at the end of the message indicating any change from present conditions. Abbreviations used are:

AT	At	NOSIG	No significant change
BECMG	Becoming	NSW	No significant weather
FM	From	TEMPO	Temporarily
GRADU	Gradually	TL	Until

e.g. (a) BECMG FM1100 250/35G50 means: becoming from 1100 wind 250 degrees 35 knots max 50 knots; (b) TEMPO FM0630 TL0830 3000 SHRA means: temporarily, from 0630 to 0830, visibility 3000 metres, showers of rain.

TAF (Terminal Airfield Forecast)

This uses essentially the same code as a METAR. There are additional groups to show the period of the forecast, e.g.

EGGW   1019   23010KT   9999   SCT010   BKN018   BECMG1114   6000   RA   BKN012
TEMPO1418   2000   DZ   OVC004   FM18   30020G30KT   9999   −SHRA   BKN015CB

Meaning

EGGW	Airfield:	Luton
1019	Time validity:	1000 to 1900 GMT
23010KT	Wind:	230 degrees 10 knots
9999	Visibility:	10 kilometres or more
SCT010	Cloud:	scattered at 1000 feet
BKN 018		broken at 1800 feet
BECMG 1114		becoming from 1100 to 1400
6000	Visibility:	6 kilometres
RA	Weather:	moderate rain
BKN012		broken 1200 feet
TEMPO 1418		temporarily 1400 to 1800
2000	Visibility:	2000 metres
DZ	Weather:	moderate drizzle
OVC004		overcast 400 feet
FM18		from 1800
30020G30KT	Wind:	300 degress 20 knots max 30 knots
9999	Visibility:	10 kilometres or more
−SHRA	Weather:	slight rain showers
BKN015CB	Cloud:	broken cumulonimbus 1500 feet

Appendix 3

SHANNON VOLMET   3413 kHz (night)
                                5505 kHz 24 hours
                                8957 kHz 24 hours
                              13264 kHz (day time)

The radio set must be able to receive single-sideband transmissions (the upper sideband is normally used).

Time	Contents	Stations
H + 00	TAF	Brussels, Hamburg
	METAR	Brussels, Hamburg, Frankfurt, Cologne-Bonn Dusseldorf, Munich
H + 05	TAF	London/Heathrow, Prestwick, Shannon
	METAR	London/Heathrow, Shannon, Prestwick London/Gatwick, Amsterdam, Manchester
H + 10	METAR	Copenhagen/Kastrup, Stockholm/Arlanda Gothenburg/Landvetter, Bergen, Oslo/Fornebu Helsinki/Vantaa, Dublin, Barcelona
H + 15	TAF	Madrid/Barajas, Lisbon, Paris/Orly
	METAR	Madrid/Barajas, Lisbon, Santa Maria Paris/Orly, Paris/Charles de Gaulle Lyon/Satolas
H + 20	TAF	Rome/Fiumicino, Milan/Malpensa
	METAR	Rome/Fiumicino, Milan/Malpensa, Zurich Geneva, Turin, Keflavik
H + 30	TAF	Frankfurt, Cologne-Bonn
	METAR	Brussels, Frankfurt, Cologne-Bonn, Dusseldorf, Munich
H + 35	TAF	London/Gatwick, Amsterdam
	METAR	London/Heathrow, Shannon, Prestwick, London/Gatwick, Amsterdam, Manchester
H + 40	METAR	Copenhagen/Kastrup, Stockholm/Arlanda, Gothenburg/Landvetter, Bergen, Oslo/Fornebu, Dublin, Barcelona
H + 45	TAF	Santa Maria, Athens, Paris/Charles de Gaulle
	METAR	Madrid/Barajas, Lisbon, Santa Maria, Paris/Orly, Paris/Charles de Gaulle, Lyon/Satolas

| H + 50 | TAF | Zurich, Geneva |
| | METAR | Rome/Fiumicino, Milan/Malpensa, Zurich, Geneva, Turin, Keflavik |

SIGMET MESSAGES at H + 0, H + 10, H + 20, H + 30, H + 40, H + 50 when necessary.
METARS may be repeated if the full five-minute segment is not filled.

RAF VOLMET LIST (August 1999, liable to changes)
(issued in six-minute segments)

H + 00 and H + 30   Belfast Aldergrove, Manchester, Prestwick, London Stansted, Bardufoss, Bodo, Oslo, Gibraltar, Horta

H + 06 and H + 36   Benson, Brize Norton, Brueggen, Geilenkirchen, Hannover, Lyneham, Northolt, Odiham, Lyneham (repeated)

H + 12 and H + 42   Coltishall, Cranwell, Leeming, Leuchars, Lossiemouth, Marham, St Mawgan, Waddington, Kinloss

H + 18 and H + 48   Ascension, Bahrain, Brize Norton, Gatow, Keflavik, Mombasa, Nairobi, Montevideo, Rio de Janeiro

H + 24 and H + 54   Adana, Akrotiri, Ancona, Aviano, Gioa del Colle, Rimini, Rome Ciampino, Skopje, Split, Brindisi, Waddington

Certain stations may close down at night, at weekends and over Bank Holidays.

LONDON VOLMET BROADCASTS (VHF)
These are half-hourly METAR reports giving surface wind, weather, visibility*, RVR if applicable, cloud*, temperature, dew point, QNH.
*or CAVOK

Additional items included if applicable are:
recent weather, windshear, TREND, runway contamination.

The word SNOCLO will be added to the end of the report when the aerodrome is unusable for take off or landing due to heavy snow on the runways or runway snow clearance.

LONDON VOLMET (MAIN) 135.375 MHz
Amsterdam
Brussels
Dublin
Glasgow
LONDON/Gatwick
LONDON/Heathrow
LONDON/Stansted
Manchester
PARIS/Charles de Gaulle

LONDON VOLMET (SOUTH) 128.600 MHz
Birmingham
Bournemouth
Bristol

	Cardiff
	Jersey
	Luton
	Norwich
	SOUTHAMPTON/Eastleigh
	Southend
LONDON VOLMET (NORTH) 126.600 MHz	Blackpool
	East Midlands
	Leeds-Bradford
	Liverpool
	LONDON/Gatwick
	Manchester
	Newcastle
	ISLE OF MAN/Ronaldsway
	Teesside
SCOTTISH VOLMET 125.725 MHz	Aberdeen/Dyce
	BELFAST/Aldergrove
	Edinburgh
	Glasgow
	Inverness
	LONDON/Heathrow
	Prestwick
	Stornoway
	Sumburgh

More comprehensive details are now available in a free booklet entitled *GET MET: Aviation Weather Services* issued by the Met. Office and the Civil Aviation Authority in association with the British Aviation Insurance Group. For copies, contact The Met. Office, Sutton House, London Road, Bracknell, Berkshire RG12 2SY.

Acknowledgement
This appendix has used data supplied by the CAA and Meteorological Office.

Further Reading

M. J. Bader, G. S. Forbes, J. R. Grant, R. B. E. Lilley and A. J. Waters, *Images in Weather Forecasting*, 1995 Cambridge University Press.
Techniques for using satellite and radar images.

Robert N. Buck, *Weather Flying* (4th ed.), 1998 McGraw-Hill.
Very readable, practical book by a highly experienced pilot.

W. J. Burroughs, Bob Crowder, Ted Robartson, Eleanor Vallier-Talbot and Richard Whittaker, *Weather: The Ultimate Guide to the Elements*, 1996 HarperCollins.
Exceptionally good illustrations.

Brian Cosgrove, *Pilot's Weather*, 1999 Airlife Publishing.
Many colour photos and diagrams.

Ingrid Holford, *Guinness Book of Weather Facts and Feats*, 1977 Guinness Superlatives Ltd.
Records and extreme weather conditions.

Robert A. Houze Jr, *Cloud Dynamics*, 1993 Academic Press.
Professional work with a number of equations.

F. H. Ludlam, *Clouds and Storms*, 1980 Pennsylvania State University.
University standard, very comprehensive, with equations.

Met. Office, *Meteorological Glossary* (6th ed.), 1991 HMSO.
Extremely useful.

Met. Office, *Handbook of Aviation Meteorology* (3rd ed.), 1994 HMSO.
The prime manual on which many exams are based.

Denis Pagen, *Understanding the Sky*, 1992 Denis Pagen.
Mainly for hang-glider pilots. Many diagrams.

Roger Pielke Jr and Roger A. Pielke Sr, *Hurricanes*, 1997 John Wiley & Sons.
Nature and impact of hurricanes on society.

Derek Piggott, *Understanding Flying Weather* (2nd ed.) 1996 A & C Black.
Good, simple introduction to meteorology for learners in gliding.

Elmar R. Reiter, *Jet-stream Meteorology*, English translation 1963 University of Chicago.
Comprehensive details of early research on jet-streams.

R. S. Scorer, *Environmental Aerodynamics*, 1978 Ellis Horwood.
Mathematical approach to airflow and many meteorological subjects.

R. S. Scorer, *Cloud Investigation by Satellite*, 1986 Ellis Horwood.
Fascinating series of satellite pictures with explanations.

R. S. Scorer, *Satellite as Microscope*, 1990 Ellis Horwood.
Very detailed account of satellite pictures.

R. S. Scorer, *Dynamics of Meteorology and Climate*, 1997 John Wiley & Sons.
Very comprehensive, partly mathematical treatment.

Richard Scorer and Arjen Verkaik, *Spacious Skies*, 1989 David & Charles.
Exceptional set of pictures and expert text.

John E. Simpson, *Gravity Currents*, 1987 Ellis Horwood.
Includes many weather phenomena such as sea-breezes.

John E. Simpson, *Sea Breeze and Local Wind*, 1994 Cambridge University Press.
Useful account of many local winds.

C. E. Wallington, *Meteorology for Glider Pilots* (3rd ed.), 1977 John Murray.
The original book on the subject.

A.nn Welch, *Pilots' Weather*, 1973 John Murray.
Straightforward and uncomplicated flying manual.

World Meteorological Organisation, *Handbook of Meteorological Forecasting for Soaring Flight*, Technical Note no. 158 (2nd ed.), 1993 WMO, no. 495.
A practical text-book for forecasters prepared by the OSTIV Secretariat, Oberpfaffenhofen, Germany.

Index